Springer Series in Design and Innovation

Volume 3

Springer Series in Design and Innovation (SSDI) publishes books on innovation and the latest developments in the fields of Product Design, Interior Design and Communication Design, with particular emphasis on technological and formal innovation, and on the application of digital technologies and new materials. The series explores all aspects of design, e.g. Human-Centered Design/User Experience, Service Design, and Design Thinking, which provide transversal and innovative approaches oriented on the involvement of people throughout the design development process. In addition, it covers emerging areas of research that may represent essential opportunities for economic and social development.

In fields ranging from the humanities to engineering and architecture, design is increasingly being recognized as a key means of bringing ideas to the market by transforming them into user-friendly and appealing products or services. Moreover, it provides a variety of methodologies, tools and techniques that can be used at different stages of the innovation process to enhance the value of new products and services.

The series' scope includes monographs, professional books, advanced textbooks, selected contributions from specialized conferences and workshops, and outstanding Ph.D. theses.

Keywords: Product and System Innovation; Product design; Interior design; Communication Design; Human-Centered Design/User Experience; Service Design; Design Thinking; Digital Innovation; Innovation of Materials.

How to submit proposals

Proposals must include: title, keywords, presentation (max 10,000 characters), table of contents, chapter abstracts, editors'/authors' CV.

In case of proceedings, chairmen/editors are requested to submit the link to conference website (incl. relevant information such as committee members, topics, key dates, keynote speakers, information about the reviewing process, etc.), and approx. number of papers.

Proposals must be sent to: series editor Prof. Francesca Tosi (francesca.tosi@unifi.it) and/or publishing editor Mr. Pierpaolo Riva (pierpaolo.riva@springer.com).

More information about this series at http://www.springer.com/series/16270

Isabel Dias Cabral · António Pedro Souto · Linda Worbin

Dynamic Light Filters

Smart Materials Applied to Textile Design

Springer

Isabel Dias Cabral
2C2T-Centre for Textile Science
and Technology
University of Minho
Guimarães, Portugal

António Pedro Souto
Department of Textile Engineering
2C2T-Centre for Textile Science
and Technology
University of Minho
Guimarães, Portugal

Linda Worbin
Material and Innovation Developer
IKEA
Älmhult, Sweden

ISSN 2661-8184 ISSN 2661-8192 (electronic)
Springer Series in Design and Innovation
ISBN 978-3-030-39528-5 ISBN 978-3-030-39529-2 (eBook)
https://doi.org/10.1007/978-3-030-39529-2

This Springer imprint is published by the registered company Springer Nature Switzerland AG
The registered company address is: Gewerbestrasse 11, 6330 Cham, Switzerland

Preface

Emergent materials and technologies originated by scientific and technological advances have been providing new means to extend textiles' conventional qualities. Smart materials are able to sense and react reversibly to an external stimulus and when integrated in textiles, embed them with intrinsic dynamic and interactive behavior.

Smart textiles have been creating great interest for research and development of innovation-driven applications. They enable the development of new and enhanced functionalities, expressions and performances, namely through the potential to interact with our environment, creating relationships between physical and immaterial dimensions.

Design and development of smart textiles encompass competences in diverse domains and entail new perspectives and challenges for design research and practice, given the novelty that reactive and adaptive qualities introduce.

This book provides interdisciplinary and articulated coverage on smart textiles topics and discusses original research developed on the integration of smart materials in textile substrates and textile design. The design concept subjacent to the research hinges on the interaction of textiles and light. With natural light, man has a long history of using textiles to change light intensity and tone of sunlight, using for example awnings, curtains, clothing and accessories. With artificial light, we commonly act upon the light source: switching on and off different lamps and using dimmers or other technologies to change lighting parameters.

Focused on our experience working with natural light and textiles, the research discussed in this book aimed to study and develop smart textiles that can change the incident light that passes through them—light transmittance—without acting upon the light source, thus creating Dynamic Light Filters.

The research conducted was based on knowledge and experimental practices of textile engineering and design, comprising of two approaches: (a) a materials research focused on integration processes of color change, shape memory and conductive materials in textile substrates; (b) a design research that studied dynamic qualities of textile color and shape behavior in interaction with light.

After an introduction to Smart Textiles and Dynamic Light Filters themes, this book examines fundamental concepts and knowledge in each smart material domain and respective materials research. Part I looks into dynamic color in textiles through Color Change Materials, Thermochromic textiles, their thermo-responsive behavior and respective electrical activation. Part II focuses on dynamic form through Shape Memory Alloys and textile morphological performances based on origami techniques. Part III presents practice-based design research that explores dynamic qualities of color and shape thermo-responsive textiles and their interaction with light. The research program comprises of two main experimental studies on textile behavior and dynamic light, followed by the development and discussion of three research prototypes that proposed to explore expressive possibilities of color, shape and light performances through different intensity levels of change.

This book will be of interest to academicians, researchers and practitioners inspired by smart materials and textile dynamic and interactive behavior. Application fields include engineering, design, architecture and arts. Sections that comprise of experimental work conducted on integration processes of smart materials in textile substrates are also of value to the textile industry.

The authors would like to express their gratitude to many people and entities that contributed to the preparation of this book and respective research discussed within it.

The research presented in this book was financed by FEDER funds through the Operational Programme for Competitiveness Factors—COMPETE and National Funds through FCT—Foundation for Science and Technology within the scope of the projects SFRH/BD/87196/2012 and Project UID/CTM/00264/2019.

Authors acknowledgement goes to 2C2T—Centro de Ciência e Tecnologia Têxtil (Center for Textile Science and Technology), University of Minho.

Authors acknowledge Prof. Helder Carvalho and Prof. Joana Cunha for their support during research development, as well as Prof. Fernando Moreira da Siva, Prof. Manuel José dos Santos Silva and Prof. Mário de Araújo for critical discussion of this research.

The authors also express their acknowledgment to Smart Textiles Design Lab (STDL) for the support on the prototypes development and to Prof. Olle Holmud and Magnus Sirhed for the assistance provided during the work developed in Weaving Lab of the Swedish School of Textiles, University of Borås.

For providing the opportunity to present the prototypes in a space context, authors gratitude to Ikea Museum, Smart Textiles Salon and Casa da Música (House of Music).

Authors also acknowledge OTOJAL for the development of a research prototype through rotary printing.

Thank you to Cristiano Silva for his collaboration during the development of the electronic circuits and programing of the research prototypes and to Manuel Cabral and Agostinho Vieira for the collaboration in video and photographic record of the prototypes II and III.

The authors also thank Leontina Di Cecco, Senior Editor, and Springer for the opportunity and support in the preparation of this book.

Extended gratitude to colleagues, friends and family that accompanied this journey.

Guimarães, Portugal — Isabel Dias Cabral
Guimarães, Portugal — António Pedro Souto
Älmhult, Sweden — Linda Worbin

Contents

Part II Dynamic Form in Textiles

Part III Dynamic Light Filters

Abbreviations

A	Austenite
Af	Austenite finish
As	Austenite start
CCM	Color Change Material
CIE	Commission Internationale de l'Eclairage
CIELAB	CIE L*a*b* system
CO	Cotton
CT	Conductive Thread
M	Martensite
Mf	Martensite finish
Ms	Martensite start
PE	Polyethylene
PES	Polyester
SMA	Shape Memory Alloy
SME	Shape Memory Effect
SMM	Shape Memory Material
TC	Thermochromic

Chapter 1
Introduction

Textiles have been a prominent interface with our environment and a prime example of innovation-driven developments. Emerging sensing and reacting capabilities are imparted to textiles by means of smart materials and enabling conductive and electronic technologies. Introducing new qualities to textile conventional functions and expressions, smart textiles have been opening up vast potential for the creation of adaptive and reactive surfaces.

Through the perspective of seamless integration of technology into our environment, smart textiles are able to involve physical and immaterial dimensions, promoting added functionality and interaction between people and their surroundings. In interior spaces, variation of artificial light intensity and tone are commonly achieved by acting upon the light source. Smart textile responsive behavior presents the possibility to change the light that passes through them—light transmittance—through the reversible change of their properties, performing as Dynamic Light Filters and allowing the design of dynamic lighting scenarios.

Design and development of smart textiles encompasses competences in diverse domains and presents new challenges: a technical and creative understanding of the materials' properties, behavior and processing possibilities as a materials system are required as well as consideration of their active qualities. In parallel, smart textiles demand a rethinking of conventional design variables and methodologies, given the novelty that dynamic and interactive dimensions introduce.

This book provides interdisciplinary and articulated coverage on smart textiles topics and discusses original research developed on integration of smart materials in textile substrates based on systematic processes, dynamic qualities of color and shape thermo-responsive textiles, and design potential of smart textile behavior to dynamically filter light.

After this chapter introduction to Smart Textiles and Dynamic Light Filters themes, this book examines fundamental concepts and knowledge in each smart material domain and respective research. Part I looks into dynamic color in textiles through Color Change Materials, their thermo-responsive behavior and respective electrical activation. Part II focuses on dynamic form through Shape Memory

I. D. Cabral et al., *Dynamic Light Filters*, Springer Series in Design and Innovation 3, https://doi.org/10.1007/978-3-030-39529-2_1

Alloys and textile morphological performances based on origami techniques. Part III presents design research that explores innovative perspectives on smart textiles dynamic qualities, consideration of design variables and possibilities of smart textiles and light interaction to create responsive lighting environments.

1.1 Smart Textiles

Textiles have been a prominent interface with our environment and a prime example of innovation-driven developments. Among diverse possibilities, textiles have been applied for sun shading and transformation of wind into motion energy. They provide enhanced qualities for high-performance applications, such as lighter and stronger architectural structures, protective professional clothing and biocompatible implants. They are at the basis of technological advances, namely materials processing and computing-based technologies and are able to sense and dynamically react to external stimuli (McQuaid 2005; Quinn 2010; Kettley 2016).

Emerging sensing and reacting capabilities are imparted to textiles by means of smart materials and enabling conductive and electronic technologies (Koncar 2016). Smart materials exhibit sensitive-stimulus capability that triggers a reversible response, sensing inputs include thermal, mechanical, electrical and magnetic energy, among others (Tao 2001; Langenhove 2015).

When smart materials are integrated in textile substrates, they embed them with intrinsic dynamic and interactive behavior, thus presenting new qualities to conventional textiles' functions and expressions (Worbin 2010; Kirstein 2013). Dynamic quality refers to the textile ability of property change and interactive quality is due to the changes in response to a sensed stimulus. For example, thermochromic (TC) textiles change color upon temperature variation and, through the same stimulus, shape memory textiles can change shape.

Moreover, smart textiles can also combine data processing, communication and power supply functions, enabled by increased progress of textile-based conductive materials and electronics miniaturization that impart textiles with electronic and computation capabilities. This character is commonly classified as very smart, describing the textile ability of sensing, reacting and adapting (Tao 2001).

Driven by scientific and technological developments, smart materials have been gaining increased attention for textile research and applications, namely in biomedical, protection, fashion and communication fields (Merati 2018). As on-going research into new materials and enabling technologies is moving forward, creative approaches are also exploring and unfolding new textile futures as interactive surfaces of our daily life.

Characterized by interdisciplinary convergence, research and development of smart textiles combine knowledge and competence from materials science, electronics, textile engineering, textile design, interaction design, etc. Towards performance feasibility of textiles, collaborative research seeks to embed and explore functional and interactive dimensions in an unobtrusive way, so that textile intrinsic properties or designed expressions are not significantly compromised, for example stretching, recovery, draping, shearing or handling (Zysset et al. 2013; Stoppa and Chiolerio 2014; Kumar and Vigneswaran 2016; Yilmaz 2019).

The perspective of seamless technology integration is transversal to diverse fields. The 'ubiquitous computing' concept, introduced by Marc Weiser in the late 1980s, envisioned that "The most profound technologies are those that disappear. They weave themselves into the fabric of everyday life until they are indistinguishable from it" (Weiser 1991). Whereas technology is embedded into our environment (Ambient Intelligence), occurring through continuous interconnections (Pervasive Computing) or regarding to clothing and accessories (Wearable Electronics), the main aims rely on built-in active performances able to enhance welfare, intellect, creativity and communication, also stimulating sensory and emotional fulfilment (Baurley 2005; Ko et al. 2005; Aarts and Wichert 2009; Schneegass and Amft 2017).

Textiles as a pervasive and soft interface for new technologies and interactive performances, challenge our concept of textiles, commonly associated with passive functionality, as well as the way we may use or interact with them (Frumkin and Weiss 2012; Pailes-Friedman 2016). In addition to encompassing competences in diverse domains, smart textiles also add new parameters to design research and practice. Design reactive and adaptive qualities in textiles requires technical and creative understanding of the materials behavior, process possibilities in respect to smart textiles' physical materialization and their behavior framework, as well as a rethinking of conventional design variables and methodologies given the novelty that dynamic and interactive dimensions introduce (Worbin 2010; Vallgårda 2014; Mossé 2016). Time and movement as emergent dimensions of textile design entail new perspectives in respect to how the textile changes and what it expresses, as well as how their behavior can be designed.

1.2 Dynamic Light Filters

Through the perspective of seamless integration of technology into our environment, smart textiles are able to involve physical and immaterial dimensions, promoting added functionality and interaction between individuals and their surroundings (Bonnemaison and Macy 2007; Benitez 2016). In interior spaces, variation of artificial light intensity and tone are commonly achieved by acting upon the light source. The emergent sensing and reacting behavior of smart textiles present the potential to change the incident light that passes through them—light transmittance (Yot 2011)—to design dynamic lighting scenarios, performing as *Dynamic Light Filters*.

The interaction between textiles and light is designed through the textile properties and dynamic qualities. The topics and research presented in this book explore smart textiles and light transmittance dynamic variables through textile color and shape changing behavior.

The emergence of Color Change Materials (CCM) allowed the introduction of chromatic dynamic qualities to textiles, a behavior based on the variation of the substances microstructure or electronic state, affecting their optical characteristics—absorptance, reflectance, scattering or transmittance (Addington and Schodek 2005; Bamfield and Hutchings 2010).

TC leuco dyes are a class of CCMs that through thermal stimulus change their visual characteristics reversibly. They perform color change by fading away above an activation temperature and returning to the predefined color below it. Able to be mixed with conventional pigments, instead of a colorless effect, heat variation results in changes from one color to another. As dark colors absorb a greater intensity of the visible light spectrum than lighter colors (Descottes and Ramos 2011), TC textiles can affect the light that pass through them, when they are below or above their activation temperature.

The design concept inherent to the TC materials research presented in this book aimed to transform similar light intensities and tones to heterogeneous, as well as the inverse. In this sense, experimental work with TC pigments concerned with the development of systematic processes to create textiles that change color according to predefined ratios.

Thermo-responsive Shape Memory Alloys (SMAs) are metal compounds capable to change from a temporary to a pre-programmed shape, upon a heat stimulus (Otsuka and Wayman 1998). By displaying interactive motion and force, these materials challenge new perspectives towards innovative and reliable applications, as well as revealing potential to supersede other actuators (Lawson 2016). Their unique properties include stimulus-sensitive with a silent kinetic response and high actuation forces, as well as lightweight and biocompatible options (Bengisu and Ferrara 2018; Mehta and Gupta 2019). As different amounts of textile layers interfere in light transmittance, depending on the light absorbed, it is possible to develop shape memory textiles that perform morphological variations with variable layer numbers. This requirement was studied through geometric structures based on origami techniques, an ancient Japanese art of folding paper.

Defined by the ability to conduct electrical current, conductive materials are a key topic in smart textiles field, as they enable the transfer of energy or data and impart textiles with electronic functionalities (Storey 2009; Eichhoff et al. 2013). Working with thermo-responsive textiles, conductive materials enable the electrical activation of color and shape behavior through resistive heating, according to their properties, integration in the substrate and thermal conductivity of the textile structure. In the research conducted, integration of conductive materials in textiles aimed to produce thermal variation of the TC pigments. SMAs are conductive materials and shape change activation was performed through their resistive heating properties.

Smart textiles design comprises of a synergetic relationship of the textiles tangible and immaterial aspects. Considering time and movement as critical variables that smart textiles introduce, they also entail new perspectives and alternative methods to understand how to work with temporal forms in textiles (Redström 2010; Mossé 2016). With thermo-responsive textiles, heat is a main dynamic variable to explore design possibilities of textile color and shape behavior and, in this research context, the creation of dynamic lighting.

This book provides interdisciplinary and articulated coverage on smart textiles topic and discusses original research developed on integration of smart materials in textile substrates based on systematic processes, dynamic qualities of color and shape thermo-responsive textiles, as well as design potential of smart textile behavior to dynamically filter light.

Part I and Part II look into dynamic color and form in textiles, respectively, and from a material research perspective, Chap. 3 presents the study and development of systematic processes of paste recipe formulation with TC and conventional pigments to screen print textiles with defined color change ratios from similar to different with temperature increase and the inverse; Chap. 4 explain the study of integration processes of metal-based conductive materials in woven substrates and the analysis of their electrical and expressive qualities for resistive heating activation of thermo-responsive textiles; and Chap. 6 discusses the study and development of a workflow setup to design and manufacture shape memory woven textiles in which dynamic behavior achieves predefined geometric morphologies, performing layer number variation.

From a design perspective, Part III presents the practice-based design research developed to explore dynamic qualities of color and shape thermo-responsive textiles and their interaction with light. The research program comprises of two main experimental studies on textile behavior and dynamic light, followed by the development and discussion of three research prototypes. Design and presentation of the prototypes propose to study dynamic qualities of thermo-responsive textiles behavior based on selected design variables and discuss expressive possibilities of color, shape and light performances through different intensity levels of change.

References

Aarts E, Wichert R (2009) Ambient intelligence. In: Bullinger H-J (ed) Technology guide: principles—applications—trends. Springer, Berlin Heidelberg, Berlin, Heidelberg

Addington M, Schodek D (2005) Smart materials and new technologies. Architectural Press, Amsterdam

Bamfield P, Hutchings MG (2010) Chromic phenomena: technological applications of colour chemistry. Royal Society of Chemistry, Cambridge

Baurley S (2005) Interaction design in smart textiles clothing and applications. In: Tao X (ed) Wearable Electronics and Photonics. Woodhead Publishing, Cambridge

Bengisu M, Ferrara M (2018) Materials that move: smart materials, intelligent design. Springer International Publishing, Cham

Benitez M (2016) Sensorial space: responsive interiors through smart textiles. In: Schneiderman D, Winton A (eds) Textile technology and design: from interior space to outer space. Bloomsbury Academic, New York

Bonnemaison S, Macy C (2007) Introduction. In: Bonnemaison S, Macy C (eds) Responsive textile environments. TUNS Press, Halifax, N.S.

Descottes H, Ramos CE (2011) Architectural lighting: designing with light and space. Princeton Architectural Press, New York

Eichhoff J, Hehl A, Jockenhoevel S, Gries T (2013) Textile fabrication technologies for embedding electronic functions into fibres, yarns and fabrics. In: Kirstein T (ed) Multidisciplinary know-how for smart textiles developers. Woodhead Publishing, Oxford
Frumkin S, Weiss M (2012) Fabrics and new product development. In: Horne L (ed) New product development in textiles: innovation and production. Woodhead Publishing Ltd, Oxford
Kettley S (2016) Designing with smart textiles. Fairchild Books Bloomsbury, London
Kirstein T (2013) The future of smart-textiles development: new enabling technologies, commercialization and market trends. In: Kirstein T (ed) Multidisciplinary know-how for smart textiles developers. Woodhead Publishing, Oxford
Ko F, Aufy A, Lam H, Macdiarmid A (2005) Electrostatically generated nanofibres for wearable electronics. In: Tao X (ed) Wearable electronics and photonics. Woodhead Publishing, Cambridge
Koncar V (2016) Introduction to smart textiles and their applications. In: Koncar V (ed) Smart textiles and their applications. Woodhead Publishing, Amsterdam
Kumar L, Vigneswaran C (2016) Electronics in textiles and clothing: design, products and applications. CRC Press, Taylor & Francis Group, Boca Raton, FL
Langenhove LV (2015) Smart textiles: past, present, and future. In: Tao X (ed) Handbook of smart textiles. Springer, Singapore
Lawson BC (2016) Shape memory alloys, muscle wires and robotics. S.I., CreateSpace
Mcquaid M (2005) Extreme textiles: designing for high performance. Smithsonian Cooper-Hewitt, National Design Museum, New York
Mehta K, Gupta K (2019) Fabrication and processing of shape memory alloys. Springer, Cham
Merati AA (2018) Application of stimuli-sensitive materials in smart textiles. In: Shahid-Ul-Islam, Butola BS (eds) Advanced textile engineering materials. Scrinever Publishing, Beverly
Mossé A (2016) Self-actuated textiles, interconnectivity, and the design of the home as a more sustainable timescape. In: Schneiderman D, Winton A (eds) Textile technology and design: from interior space to outer space. Bloomsbury Academic, New York
Otsuka K, Wayman CM (1998) Shape memory materials. Cambridge University Press, Cambridge
Pailes-Friedman R (2016) Smart textiles for designers: inventing the future of fabrics. Laurence King Publishing, London
Quinn B (2010) Textile futures: fashion, design and technology. Berg, Oxford
Redström J (2010) On technology as material in design. In: Redström M, Redström J, Mazé R (eds) IT+Textiles. Borås, The Interactive Institute and the Swedish School of Textiles
Schneegass S, Amft O (2017) Introduction to smart textiles. In: Schneegass S, Amft O (eds) Smart textiles: fundamentals, design, and interaction. Springer International Publishing, Cham
Stoppa M, Chiolerio A (2014) Wearable electronics and smart textiles: a critical review. Sensors 14:11957. https://doi.org/10.3390/s140711957
Storey N (2009) Electronics: a systems approach. Pearson Education Limited, Harlow
Tao X (2001) Smart technology for textiles and clothing—introduction and overview. In: Tao X (ed) Smart fibres, fabrics and clothing: fundamentals and applications. Woodhead Publishing, Cambridge
Vallgårda A (2014) Giving form to computational things: developing a practice of interaction design. Pers Ubiquit Comput 18(3):577–592. https://doi.org/10.1007/s00779-013-0685-8
Weiser M (1991) The computer for the 21st century. Sci Am 265(3):66–75. https://doi.org/10.1145/329124.329126
Worbin L (2010) Designing dynamic textile patterns. Doctoral thesis, University of Borås
Yilmaz ND (2019) Introduction to smart nanotextiles. In: Yilmaz ND (ed) Smart textiles: wearable nanotechnology. Wiley Scrivener Publishing, Hoboken
Yot R (2011) Light for visual artists: understanding & using light in art & design. Laurence King, London
Zysset C, Kinkeldei T, Münzenrieder N, Tröster G, Cherenack K (2013) Fabrication technologies for the integration of thin-film electronics into smart textiles. In: Kirstein T (ed) Multidisciplinary know-how for smart textiles developers. Woodhead Publishing, Oxford

Part I
Dynamic Color in Textiles

Chapter 2
Dynamic Color

Materials with dynamic color behavior present the opportunity to embed textiles with sensing and reacting qualities. Reversible chromatic changes are imparted to textiles through the integration of Color Change Materials (CCMs), which are classified according to the stimulus that induces the changes, as thermochromic (TC) pigments are an example by changing color upon a thermal input.

Dynamic Light Filter effect can be designed through the interaction of textile color behavior and light. As dark colors absorb a greater intensity of visible light spectrum than lighter colors, TC textiles affect the light than passes through them, when they are below or above their activation temperature.

For an introduction of dynamic color in textiles, this chapter discusses light and qualitative lighting concepts and presents an overview of lighting in design, arts and architecture. Secondly, it introduces fundamental concepts of color, color theory and colorimetry. Third and fourth sections review the CCMs properties and behavior with particular focus on thermal sensitive colourants—TC—and examine developing technologies for the integration in textiles and design examples.

2.1 Light

Natural or artificial light plays an important role in human life. It ensures visibility, it affects our biological rhythms and, among various functional and aesthetic aspects, light also influences how our surroundings are perceived and experienced, for example presenting the ability to visually transform materials' textures and colors as well as space boundaries (Karlen et al. 2012).

It is through interaction with matter that light is perceived, at the same time it also makes objects and environment visible (Cuttle 2015). Visible light consists of an electromagnetic spectrum with wavelengths between approximately 380 and 740 nm and when it strikes a surface, diverse phenomena can occur (Choudhury 2014b). Opaque materials reflect light—reflectance—whereas translucent or transparent materials let

I. D. Cabral et al., *Dynamic Light Filters*, Springer Series in Design and Innovation 3, https://doi.org/10.1007/978-3-030-39529-2_2

the light pass through them—transmittance. The amount of light that is reflected or transmitted depends on the proportion absorbed (Yot 2011). Materials' properties also affect the degree of light diffusion. Matte surfaces scatter the incident light at several angles, while in the case of specular surfaces, light is reflected in a singular direction and there is no diffusion (Innes 2012).

Furthermore, interaction of light with matter is also crucial to establish color stimuli. Color of light is defined by its spectral composition, which respects to the wavelength spectrum and respective radiance. When light strikes a surface, specific light wavelengths are absorbed according to the materials' optical characteristics, changing the spectral composition of light reflected or transmitted and observed as the materials' color (Descottes and Ramos 2011; Lechner 2015).

Perception of light that is reflected from a surface—brightness—does not depend merely on its physical quantity—luminance. Perceived luminosities can be influence by context, personal judgement, relation between adjacent areas, among others (Lam 1992; Maack and Pawlik 2009). In 1957, S. S. Stevens studied the perception of light change with different magnitude levels, observing that doubling light intensity is not necessarily perceived as twice as bright. When the stimulus doubled was low, the change was perceived as more than twice as bright, whereas its increase resulted in low noticeable changes (Goldstein 2010). Comprehension of light as a physical and perceptual phenomenon is of great significance when designing lighting scenarios.

Visual perception is how observed stimuli are apprehended and interpreted, giving rise to our visual experience (Kalat 2007; Goldstein 2010). The importance of the observers' active role in this process was not initially considered in lighting design criteria. Previous focus relied on quantitative aspects of light for specific tasks (Descottes and Ramos 2011). A more comprehensive perspective on lighting has started to evolve, particularly through the work of Richard Kelly and William Lams, giving rise to what is referred to as perception-oriented or qualitative lighting design (Maack and Pawlik 2009; McAuliffe 2016).

Kelly declined the definition of light quantity and uniform illuminance as central lighting parameters and in the 1950s he differentiated three fundamental qualities of lighting: ambient light, focal glow and play of brilliance. Lam considered diverse relationships between lighting qualities with human needs for visual information and in the 1970s he established a set of criteria based on activity and biological factors (Ganslandt and Hofmann 1992; Lechner 2015).

The activity or activities performed in a space may present different lighting requirements, such as demands of color rendering and duration or frequency of the task. According to specific activity needs, Lam highlighted the importance of the type of light source, light direction, concentration and color in order to enable visual acuity and comfort (Lam 1992).

Biological needs also play a critical role in lighting design, being the major considerations examined by Lam the needs for orientation, physical security and definition of personal territory. For example visual acuity of field boundaries is crucial for space-orientation, whereas the definition of personal territory benefits from local lighting (Lam 1992). Furthermore, day and night light cycles influence human biological rhythms (circadian rhythms) and establish lighting associations for time-orientation, such as variations of light intensity, tone and direction (Karlen et al. 2012; Boyce 2014).

Qualitative lighting criteria also contemplates the importance of personal interpretation of what is observed, including attributive processes, with respect to the attribution of a meaning in response to stimuli; expectations, concerning prior experiences; and affective or emotional responses (Lam 1992; Goldstein 2010). A practical example that reflects the role of lighting expectations is the interpretation of a 'bright night' or a 'dark day', although the light levels of such a night are most probably much lower than the day referred.

Light inspires researchers and practitioners to shape and transform luminous environments. Drawn upon visual perception and optical interactions, lighting design has been evolving creatively, in scope and resources, such as enabling technologies and light emitting devices (Cuttle 2015). Although great focus is now placed on applications and control of artificial light sources, lighting effects and atmospheres have been timelessly created through interaction of light with materials, architectural volumes and atmospheric conditions.

In architecture, light is acknowledged with great significance not just in respect to architectural lighting but also in how building structures are conceptualized and developed to interact with light. Louis Kahn's and Tadao Ando's works use light as a central element, in particular through the interplay of light and shadows. Reflecting on his work, Kahn stated "Even a space intended to be dark should have just enough light from some mysterious opening to tell us how dark it really is. Each space must be defined by its structure and the character of its natural light." (Büttikger 1993).

Relationships between high and low luminosities have been used as a major creative tool to engage visual effects and create atmospheres, a concept associated with lighting sensorial qualities and character (McAuliffe 2016). James Turrells' light installations involve viewers through transformative perceptions, using static artificial light. In Afrum Pale Blue (1968), a light beam is projected into a room corner, which from a distance has a flat rectangle appearance, whereupon moving closer, a 3D cube is perceived. Two Blues installation (2008) consists of a dark room with a rectangular opening to a blue light-filled room. The dim lighting does not allow an immediate understanding of the space, just a blue rectangle is perceived. As the eyes' sensibility adapt to the light conditions, space becomes visible (Descottes and Ramos 2011; Turrell n.d.). These works reflect on the influence of light qualities and visual perception in the sense of space, considering the viewers location and observation time.

The spatial relationship between observer and light source is an essential aspect in lighting design. Besides luminous effects, the height of the light source is also associated with intimacy concepts, for example, luminaires are commonly placed closer to our bodies in private spaces than in public environments (Descottes and Ramos 2011). Variation of height hierarchies enable design ambiences that even taking place in a public domain, can still assign an intimate lighting character to a space (Du 2011).

Furthermore, lighting qualities are also intrinsically related with light diffusion degrees. Olafur Eliasson uses artificial fog to change light expression inside empty rooms, resulting in dense atmospheric effects (Eliasson n.d.). Light diffusion through fog appears to give it a tangible quality, negating the perception of spacial limits. Eliassons' atmospheric installations create light experiences with limited visibility where "the act of seeing becomes an act of feeling" (Kries and Kugler 2013).

Through functional and aesthetic perspectives, textiles have a long tradition interacting with light. Able to present varied diffusion properties, textiles are often applied to scatter the harsh light and soften it. In parallel, they also allow the transforming of luminosities and light tones, through different levels and qualities of light reflected and/or transmitted. As a result, textiles and light can be designed to provide visual and thermal comfort, enhance visual qualities for task performance, raise visual interest, create lighting atmospheres, etc. (Chilton and Lau 2015).

Examples of applications range from clothing and accessories to textiles applied in exterior or interior environments. Ancient roman velarium, was a retractable textile awning featured in coliseums to create shading for the spectators, reducing luminosity and heat (Sear 2008). Textile curtains and blinds are widely used to vary the light that comes from the exterior to create luminous environments. They are applied either for natural light during day, or blocking exterior light during night. As space dividers, textiles' translucent qualities can confer to spaces a sense of privacy and at the same time allow lighting continuity or cues. Photographic studios often use textile diffusers, changing their positioning and direction in relation to the light sources to adjust and create a particular lighting setup.

Contemporary shading systems have been a growing interdisciplinary subject focused on the design potential to interact with our environment through innovative and expressive materials, processes and performances. For both exterior and interior applications, new structural and interactive possibilities are explored to design lighting ambiences. Gerber Architekten developed a tensile textile facade in King Fahad National Library that is able to reveal and conceal light through a pattern of rhomboid textile awnings, while also aesthetically transforming the building facade (Gerber n.d.). In this example, interior lighting is defined through the interaction between natural light and the static textile structure.

In addition, increased attention in shading systems includes approaches with dynamic interactions between the environment and the tangible surfaces. In Al Bahar Towers, Aedas architects group created a facade composed by triangular polymeric panels that mechanically change their geometry though computer-system control. The variation between closed and opened areas in the facade is programmed according to weather conditions and enables a reduction of heat gains in the towers, also transforming interior lighting (Cruz 2016).

Flectofin, developed at the German Institute of Building Structures and Structural Design (ITKE) of the Stuttgart University, researches possibilities to create dynamic shading based on materials performance rather than mechanical assemblies. Inspired by the deformable behavior of Strelitzia flower, flectofin is a hingeless flapping system made of glass fibers and thermoset resin. Through the heat-responsive bending of the system spine, the direction of two flaps is shifted. By changing spatial orientation, the translucent flaps perform a visual organic open and close movement, producing self-adaptable sun-shading (Brownell 2017).

Within a myriad of applications, the interplay of textiles and light is designed through textile properties and dynamic qualities. Emergent possibilities of adaptive and reactive behavior of smart textiles open up a vast potential to create lighting scenarios that can actively interact with environmental stimuli, resulting in dynamic transformations of the textiles and light. In this book, light transmittance variations are studied through the dynamic behavior of color and shape changing textiles. In the practice-based design research presented in Chap. 7, textile performances and expressions were studied with artificial light, a medium in which qualities are commonly altered by acting upon the light source.

2.2 Color

Throughout history, mans' great interest and study of color is transversal to the arts, design, technology and science. In textiles, color is one of the main visual qualities and thus, a crucial variable in design and engineering research and practice.

Concerning diverse purposes such as aesthetic, communication, symbolic and functional, color is imparted to textiles by mean of colorant substances, namely dyes and pigments (Nassau 1998). Dyes are soluble compounds, able to migrate into textile fibers, while pigments are insoluble, thus requiring to be dispersed in a binder in order to be fixed onto the textile surface (Das 2009; Mahapatra 2016).

The use of color in textiles is intrinsically related with colorant availability and the development of application methods. Early uses of colourants depended on vegetable, animal and mineral resources, defining the possibilities of color palettes and also denoting value attributes. For example, the color purple in Greco-Roman times, known as tyrian or imperial purple, was a symbol of status due to the labor process involved, namely the quantity of murex shellfish required to extract reasonable amounts of dye. Also, it did not show a fading effect when exposed to light (Eastaugh et al. 2004; Abel 2012).

Extending the chromatics obtained directly from nature, ancient Egyptians' techniques are considered at the basis of the first synthetic inorganic pigments of which Egyptian blue pigment is an example. Although, it was only until the discovery of mauveine by William Perkin in 1856, that the synthetic dye industry was established and synthetic organic colourants for textiles underwent increased progress (Aspland 1998; Abel 2012).

Furthermore, dyeing and printing processes were developed over the centuries, such as dyeing techniques with mordants, batik, shibori, block and screen printing, which had also undergone accelerated developments during and after the industrial revolution (Clarke 2011; Briggs-Goode 2013). Innovation in science and technology domains has enabled the introduction of new colorant classes and development of new technologies to integrate and explore color in textiles, also accompanied by the establishment of regulations that include environmental considerations and quality control (Hidefi 2012).

The emergence of color change materials allowed the introduction of chromatic dynamic qualities to textiles by sensing and reacting reversibly to a stimulus, such as the TC leuco dyes that respond to heat (Schwartz 2002). The temporal and interactive dimensions of colors with these materials presented a shift in relation to traditional textiles and opened up the potential to imbue textiles with new expressions, functions and interactions. At the same time, chromic textiles development also creates new demands in respect to different complexity levels of knowledge and competences in diverse domains, entailing new perspectives and challenges for design (Baurley 2004; Worbin 2010). Design research in this area highlight the importance of experimental exploration of the materials' behavior, processes and technologies to integrate them in textiles and design methods to deeper research their potential (Worbin 2010; Robertson 2011).

To explore dynamic color in textile design and research processes to develop TC textiles that change color according to predefined ratios, it is also crucial to understand the foundations of color theory and practice, contextualized as a phenomenon that involves interaction of light with materials, detected by the eye and interpreted in the brain (Lennie 2003; Dickinson 2011).

The human visual system is able to detect and process a portion of the electromagnetic spectrum with wavelengths between approximately 380 and 740 nm, correspondent to visible light (Choudhury 2014b). Incident light on objects can be reflected (opaque materials) or transmitted (translucent or transparent materials), depending on the selective absorption of light wavelengths (Yot 2011; Kuehni 2012). The color stimulus is thus defined by the radiant energy of the wavelengths range reflected from the materials into the eye, which depends on the spectral composition of the incident light and materials' optical characteristics (Tilley 2000; Descottes and Ramos 2011). With TC leuco dyes, temperature variation changes the substances' molecular arrangement affecting their spectral reflectance, thus color change is perceived (Bamfield and Hutchings 2010).

When colors are combined, different phenomena occur in regards to the medium involved. Mixing light beams of different colors result in the wavelength addition of each light beam, a process called additive color mixing. Subtractive color mixtures respect to the combination of dyes or pigments, where each color added causes more light wavelengths to be absorbed and consequently less light is reflected into the eye (Hanson 2012; Choudhury 2014b; Cassidy 2018). Therefore, subtractive mixtures are darker than their individual components and the inverse occurs with additive mixtures.

According to TC pigments' intrinsic behavior from colored to colorless when heated, TC mixtures are darker below the pigments' activation temperature and lighter above. As darker colors absorb a vast intensity of the visible light spectrum than lighter colors, incident light that passes through a TC textile can be transformed in light transmittance tone and intensity parameters, phenomena focused in this research under the concept *Dynamic Light Filters*.

Although perception of color concerns the interaction of light and matter, ultimately it depends on how the human visual system perceives it. Physiological processes of color vision were predicted by two major theories, based on empirical experiments. The trichromatic theory, also known as the Young-Helmholtz theory, proposed that the human visual system presents three receptor mechanisms, sensitive to different ranges of light wavelengths. The opponent-process theory, also known as the Hering Opponent Color theory, described color vision in terms of three pairs of opposed chromatic responses (May 2007; Goldstein 2010; Vik 2019).

Both theories explain color vision processes that occur at different stages. When light enters the eye through the pupil, it falls upon the retina at the back of the eye. The retina has two types of photoreceptor cells known as rods and cones. Rods are sensitive to dim light, differentiating shades of black and white, while cones are sensitive to bright light and colors. As proposed by the trichromatic theory, there are three types of cone cells and each type is sensitive to a range of wavelengths: red (long wavelengths), green (middle) or blue (short). The combination of signals from the rods and cones are transmitted to the brain via the optic nerve and, as predicted by the opponent-process theory, the visual system responds to those signals through an increase or decrease of activity to indicate three pairs of opposites—red-green, yellow-blue and white-black. The combination of signals is interpreted in the brain as the color observed (Kalat 2007; Clarkson 2008; Pérez et al. 2010; Bloj and Hedrich 2016).

Based on the tristimulus nature of the human eye and light properties, each color mixture system is commonly assigned with three primary colors, meaning colors that cannot be attained by mixtures but when combined can create the wide range of colors (Hanson 2012; Cassidy 2018). Additive primary colors red, green and blue are known as RGB system. Subtractive primary colors cyan, magenta and yellow, add black (K) for printing purposes, resulting in the CMYK system. Some color systems consider blue, red and yellow as subtractive primaries and others define a three-dimensional structure based on the opponent-process colors in order to compose the color space (Bamfield and Hutchings 2010; Parraman 2012).

In addition, color perception also encompasses dynamic processes to transform the signals sensed into the experience of color (Gregory 1978; Goldstein 2010). According to Lam (1992), perception relies on a combination of sensorial data with contextual information and past experience, resulting in color being recognized or assigned under the effect of diverse factors such as color surroundings and differences between individuals.

The relationship between the sensorial information and color experience gave rise to diverse theories and principles focused on how a specific color phenomenon tends to be perceived, which commonly addresses mechanisms of chromatic adaptation (Hurlbert and Ling 2012). Color constancy respects to the ability of an object color to be recognized at some extent under different lighting conditions, through a comparison of the surroundings. For example, under a scene with blue light, the constant blue stimulus is compared and reduced to determine the objects' color (Kalat 2007; Rizzi and Bonanomi 2012).

Michel Chevreul's observations on color effects on each other resulted in the law of simultaneous contrast, which respects to the tendency of one color to induce the complementary color or opposite lightness in an adjacent or contained surface (Hurlbert and Ling 2012). This law reports to opponent processes of color vision, focusing on contextual influence of color appearance. Aside from spatial, chromatic contrast can also address temporal processing, as the color stimulus previous observed affects the visual system. For instance, the visual adaptation when one moves from a dark to a light environment and the occurrence of after-images (Krauskopf 1998; Goldstein 2010; Rizzi and Bonanomi 2012).

The perceptual phenomenon of how colors affect each other have been captured special attention of color researchers and it is a central subject in the design process, as it enables an awareness to analyze and explore color relationships in chromatic compositions (Sherin 2012). Johan Wolfgang Von Goethe focused on visual qualities of color and published his 'Theory of Colors' in 1810. Goethe studies along complementary color, contrast, shadow and proportion, were influential on further investigations, namely from Johannes Ittens' and Josef Albers' works and teaching on color, which both initiated at Bauhaus Art School (Goethe and Eastlake 2010; Dickinson 2011; Triedman 2015).

Itten explored subjective color effects and objective color principles, particularly through a color circle that he defined with complementary relations between twelve hues, presenting red, blue and yellow as subtractive primary colors, surrounded by respective secondary and tertiary hues (Feisner 2006). In his book 'The Art of Color' (1962), color relationships are also examined and demonstrated through color schemes, where besides hue, he also considers the importance of "their quantitative proportion and their degree of purity and brilliance" (Itten 1973). Itten investigated and defined color harmony and contrast based on equilibrium states, linked particularly to complementary relations, which he defined as the sum of total colors of the pigmentary mixtures that yields grey (Itten and Birren 1970).

Albers' studies and teaching on color were based on systematic experimentations, as a means to create awareness of colors' influence on each other. In 'The Interaction of Color' book (1963), Albers presents a comprehensive exploration of color interactions through practical exercises, color compositions and discussion on the principles involved, such as how simultaneous contrast can account for compositions where one color on different backgrounds can be perceived as two different colors (Albers 2013). His main aim was to provide a method for color practitioners develop "an eye for color (...) In my color book there is no new theory of color. But in it there is a way to learn to see" (Albers 1968). Itten's and Albers' works were seminal and their books are still considered the basis of color theory programs (Dickinson 2011).

Moreover, color perception also unfolds diverse assessments and interpretations depending on individual responses. Color deficiencies involve a partial loss of the color vision and interfere in the colors perceived. While aspects such as memory, experience and cultural background influence how colors are assigned to a meaning or emotional response (Feisner 2006; Goldstein 2010; Hurlbert and Ling 2012).

In regards to color specification and communication, although language is intuitive, it is also subjective and limited, given the different associations that can be undertaken by different individuals and due to the lack of precision that language affords, considering the innumerable chromatic possibilities (Albers 2013; Westland 2016).

Throughout color studies history, visual swatches have been used as a tool to communicate color more effectively, which are arranged in logical systems to specify and order appearance according to chromatic relationships. Color wheels or triangles commonly reflect relations along hues, while three-dimensional color-order systems present relations between hue (color), lightness (also named value) and chroma (or saturation), of which the Munsell color system is an example (Feisner 2006; Setchell 2012; Wardman 2018).

Colorimetry is the discipline of quantifying and physically describing human color perception, taking into account color specification, color difference evaluation and prediction of color appearance (Smith and Pokorny 2003; Goodman 2012). Numerical data is applied for a systematic color notation, also providing accurate communication (Choudhury 2010; Westland 2016).

The International Commission on Illumination (CIE—Commission Internationale de l'Éclairage) is the main standardization entity for color measurement and specified the CIE 1931 XYZ color space based on tristimulus values, which are expressed by two coordinates in a chromaticity diagram (Goodman 2012; Westland 2016; Vik 2019). CIE L*a*b* system (CIELAB) was specified in 1976 and is a color-opponent space with three coordinates: L*, a* and b*. Dimension L* corresponds to perceived lightness, representing a perfect black with $L^* = 0$ and a perfect white at $L = 100$. The coordinates a* and b* are the color-opponent dimensions with a* for red (+a*) and green (−a*) while b* for yellow (+b*) and blue (−b*). Furthermore, Chroma (C*) and hue angle (h) are defined according to L*, a* and b* values, as represented in Fig. 2.1 (Marcus 1998; Gupte 2010; Choudhury 2014b).

Color measurements are conducted under specific standards, namely illuminant and observer, where the radiant energy at each wavelength of the visible spectrum reflected from a colored surface is converted into the CIELAB color coordinates' values. The characteristics of light source are crucial parameters for color perception as well as for color measurements. CIE defined standard illuminants, in order to reproduce specific lighting conditions. D65 illuminant has a color temperature of 6504 K and represents average daylight, referring to diffuse skylight without direct

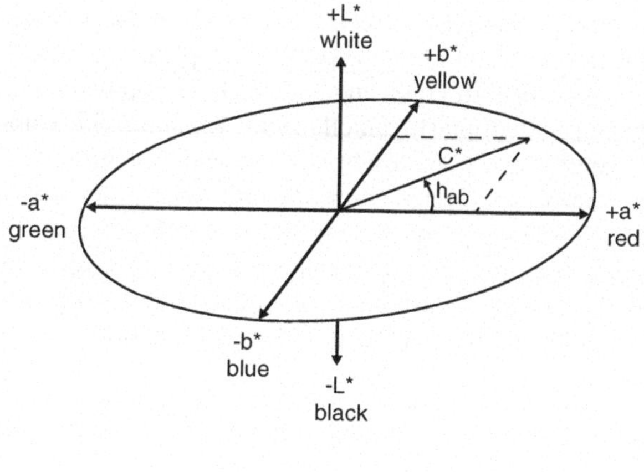

Fig. 2.1 CIE L*a*b* color space

sunlight (Gupte 2010; Wardman 2018). In regards to the eye physiology, standard observers based on a 10° viewing field have been considered a good correlation to reproduce color vision (Marcus 1998; Choudhury 2014b).

Additionally to color specification, it is also possible to quantify differences between two colors. CIELAB color difference respects to the distance between colors' location in the CIELAB space and is calculated through each color coordinate distance—dL*, da* and db*. The total color difference is expressed as dE* (Brainard 2003; Pérez et al. 2010). How different two colors can be, before it is perceptible to the human eye, can depend on diverse factors such as hue and the direction of color differences. For example, a small difference in hue between two dark color samples can be more perceptible than a higher difference in lightness between two bright color samples (Marcus 1998). Thus, tolerance limits of colors difference can guide color matching performance, but a visual assessment is also required to be considered acceptable. In the textile industry, 1.0 dE* is a common reference value (Field 1998; Gangakhedar 2010; Choudhury 2014a).

Furthermore, two colors can be perceived equal under a specific illuminant, but present differences when the illuminant changes, which is called illuminant metamerism. For a color match independent of lighting conditions, colors should have identical spectral reflectance curves, the visual profile of radiant power reflected at each wavelength of the visible light spectrum (Hunt 2004; Xin 2006; Setchell 2012).

Color reproduction involves the quantification of colorant proportions required to match a predefined color. Previously, the colourants' concentration prediction was confined to a trial and error method with visual evaluation of the results, being a time-consuming and skilled process. With colourimetric systems, it is possible to formulate colorant recipes by applying instrumental methods (Gangakhedar 2010; Choudhury 2014a).

Color formulation involves the use of a spectrophotometer, a computer and appropriate software. Calibration data of the colourants used is required, consisting on a database with the colourimetric properties measured in samples at different concentrations of each colorant. Samples' substrate and colorant application process should be equal among the concentration samples. Color formulation is calculated based on the spectral data of a pattern color and the database created with the colorant concentration samples. A recipe to reproduce the pattern color is formulated, and after the development of the respective samples, an evaluation of the results is conducted between pattern color and sample by comparison of spectral reflectance data and visual perception (Ingamells 1993; Kuehni 2012; Choudhury 2014a).

Within the complexity and subjective nature of color perception, color theory permeates perspectives to work with color systematically without excluding creative and intuitive approaches. In the research presented in this book, colorimetry and color principles also delineate possibilities to work with color change in interaction with light and contribute to an understanding of instrumental processes involved to formulate paste recipes, for color reproduction.

2.3 Color Change Materials

Color change materials are defined by the ability of changing color in response to an external stimulus. This behavior, also referred as chromic or chromogenic, is usually reversible and involves a variation of the substances microstructure or electronic state, affecting their optical properties—absorptance, reflectance, scattering or transmittance—thus, being observed as color change (Addington and Schodek 2005; Talvenma 2006; Bamfield and Hutchings 2010).

The chromic behavior is classified according to the stimulus that induces the change. The most extensively researched materials are: thermochromics, photochromics, ionochromics, electrochromics and solvatochromics, which respond to heat, light, ions, electric current and solvents, respectively. There are also chromic materials sensitive to imposed stress and/or deformations (mechanochromics), chemical changes (chemochromics), variations in magnetic field (magnetochromics), presence of pathogens (biochromics), among others (Christie 2013; Ferrara and Bengisu 2014; Viková 2019).

Color change materials present the opportunity to explore sensing and reacting qualities in design-driven innovations. Relevant applications have been found in fields such as ophthalmics, thermometry, electronics and biomedicine (Christie 2013). In textiles, this phenomenon is more recent, but increasing research and outcomes reflect the interest and potentialities of chromatic dynamic behavior.

In parallel to chromic pigments, textile color change effects can also be developed through structural colors. Pigmentary coloration refers to a photo-chemical process, whereas in structural coloration the process is photo-physical, not requiring chemical colourants. Instead, the observation of color depends on how incident light interacts with the physical structure, also being able to exhibit iridescent effect—colors observed depend on the angle of observation or illumination—and change color upon stimulus (Yuan et al. 2015; Shao et al. 2016; Yavuz et al. 2018). The research presented in this book explored dynamic color through pigmentary coloration, namely TC pigments.

Thermochromism refers to the process by which the color change effect is induced through a thermal stimulus (Day 1963; Schwartz 2002). Diverse temperature-dependent materials have been documented, however the suitability for textile application relies on encapsulated organic substances (Bamfield and Hutchings 2010; Durasevic 2016). TC microencapsulation processes date from 1970s and involve the coating of small particles or droplets of the chromic compounds to create a capsule or shell, which provides a level of protection from environmental influences (Talvenma 2006; Chowdhury et al. 2013; Salaün 2016).

The main types of TC systems used in textiles are the leuco dyes and liquid crystals (Aitken et al. 1996; Ferrara and Bengisu 2014). TC leuco dye systems exhibit a reversible color change from colored to colorless with temperature increase, a behavior that is based on chemical changes in the molecular rearrangement of the particles (Bamfield and Hutchings 2010; Salaün 2016). TC liquid crystals show different colors at a temperature range, which result from a change in their orientational structure, causing a selective reflection of specific wavelengths of light. In this case a dark background is required, in order to absorb the incident light and the reflected colors be fully observed (Talvenma 2006; Christie 2013).

Taking into consideration the objective to study light transmittance variation through color change, TC leuco dyes present the possibility to perform greater differences between darker and lighter colors than TC liquid crystals, which require a permanent dark background for the chromic effect be more obviously perceived. Thus, leuco dyes system was selected to be address in this research.

A color former, a color developer and a co-solvent compose the TC leuco dyes and the entire/complete system is contained at each microcapsule. Color former leuco dyes are electron donating and their reaction with the color developer (electron-accepting) determines the colorization of the compound, which occurs at a temperature below the melting point of the co-solvent, thus being in solid state. When the co-solvent melts at a higher temperature, it causes a dissociation of the color former and color developer, resulting in the microcapsule decolorization state (Seeboth and Lötzsch 2008; Bamfield and Hutchings 2010; Viková 2019).

The described formulation allows the selection of the TC pigment color at the lower temperature, across a range of colors and the predefinition of the pigments' activation temperature according to the co-solvent melting point, usually between −15 and 65 °C (Seeboth and Lötzsch 2013). Although TC systems can be formulated with custom-made characteristics, they are commercial available in dispersion (also called slurry) or powder forms with standard colors such as orange, red, magenta, blue and black. Activation temperature values vary according to the pigments' brand, being the most common classes approximately 7, 15, 27, 31, 41 and 47 °C.

Each TC leuco dye is confined to a variation from a color to colorless state at a defined activation temperature, although, it is possible to expand the colors palette and the chromic effect by mixing TC pigments with the same or different activation temperatures and/or by combining TC with conventional pigments, according to the defined binder type.

Mixtures of TC systems with different activation temperatures present more than two color stages with temperature variation. For example, if a TC yellow of 27 °C activation temperature is combined with a TC blue of 41 °C, the color observed in the cool state of both TC systems is the result of the colors mixing, thus green (also depending on the pigments concentration combined). When the color of the TC pigment with the lowest activation temperature fades (TC yellow), it is possible to observe the pigment color which co-solvent is in solid state—blue. Above 41 °C both pigments are colorless and the substrate color is perceived. Figure 2.2 depicts

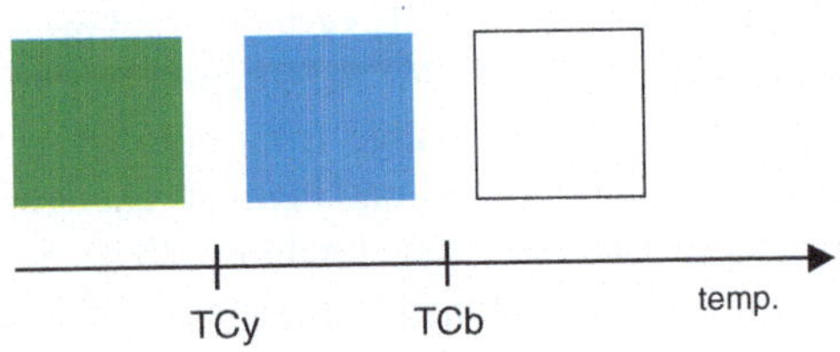

Fig. 2.2 Color change example of TC pigments combination with different activation temperatures (TC yellow and TC blue)

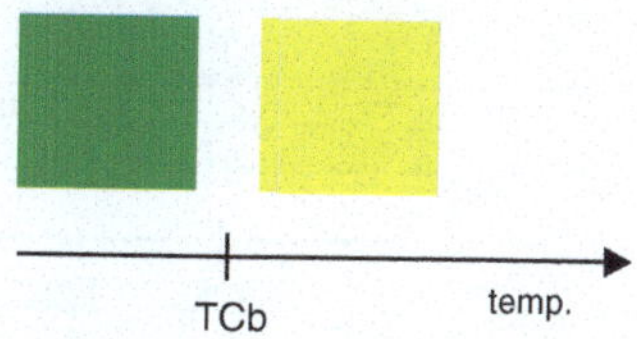

Fig. 2.3 Color change example of TC and conventional pigment combination

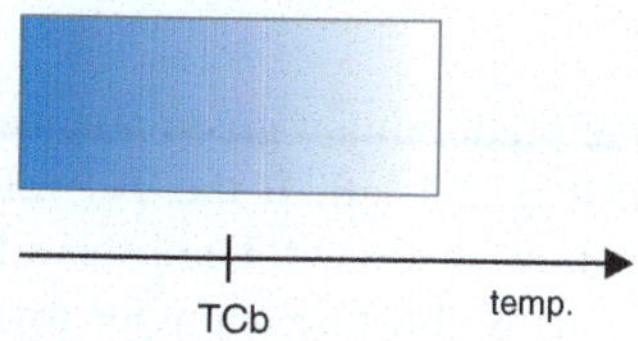

Fig. 2.4 Example of TC pigment gradual color change

the example described for temperature increase. The inverse color change process is applied for temperature decrease.

The combination of TC with conventional pigments attains a change from one color to another. If a TC pigment blue is combined with a conventional pigment yellow, below the activation temperature of the TC pigment handled, the color observed is the addition of the TC and conventional pigment colors—green. Above it, the TC pigment color fades away and the conventional pigment color is perceived—yellow, as illustrated in Fig. 2.3.

Figures 2.1 and 2.2 depict the color change effect in independent states, to describe the base behavior of TC colors and pigments combination with temperature variation. However, these materials perform a gradual behavior of color change. When TC leuco dyes are exposed to their activation temperature there is not an abrupt transition, rather, they transit through color nuances in a temperature interval (Nilsson et al. 2011), as illustrated in Fig. 2.4.

Studies on TC leuco dyes colourimetric properties have been carried out focusing on the gradual color change as a function of temperature. Heating systems with controlled temperature were used, such as a hot stage or a full cover water block, combined with a spectrophotometer for a quantitative evaluation of color at determined temperatures during heating and cooling. Results highlight that with temperature increase, color starts to change slightly before the labelled activation temperature of the pigments handled and finishes at higher temperatures, describing the gradual behavior in a temperature interval with small differences among pigments' colors and brands (Kulčar et al. 2010; Ibrahim 2012).

When TC pigments with the same activation temperature are combined, the slight differences in the pigments' activation temperature described can be observed by the appearance of new color shades in the main chromic effect (Christie et al. 2007; Nilsson et al. 2011). For example, a mixture of TC blue and TC yellow with the same labelled temperature of activation, would be expected to perform a gradual change between green and colorless on heating. However, if the TC blue presents a slightly higher activation temperature than TC yellow, a blue shade can be observed between

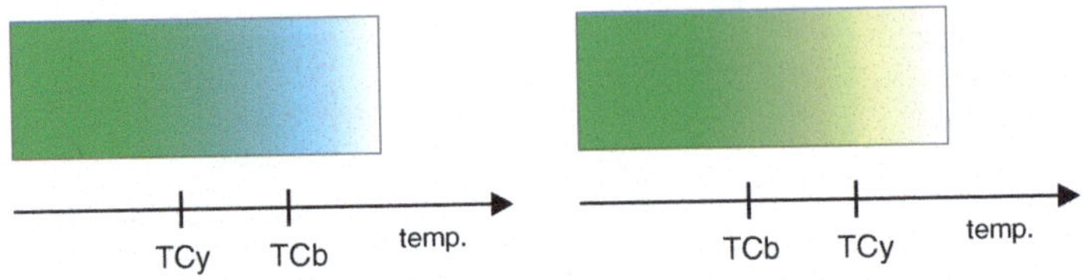

Fig. 2.5 Example of gradual color change with TC pigments combination

the predominant green nuance and the colorless state of both pigments, as depicted in Fig. 2.5 left. If the TC yellow has the slightly higher activation temperature, a yellow shade will be observed, Fig. 2.5 right.

This chromic behavior reveals the possibility to design complex color changes, where each TC pigment color added to a mixture can create an increase of color or shade in the main chromic effect (Nilsson et al. 2011). While this performance can be compared to the combination of TC pigments with labelled different activation temperatures, the color transitions attain different expressions. With TC pigments of different activation temperatures, the fading of each pigment occurs at distinct times, whereas with the similar temperature activation there is an overlap interval of the color change trajectories and does not require large temperature ranges to be observed.

Studies with color measurements on screen printed TC leuco dyes have also found that color trajectory attained during heating is different than during cooling. Commonly, a specific color lightness requires a lower temperature to be obtained in the colorizing process than when decolorizing. Thus, the gradual color change is not merely characterized as a function of temperature but also depends on the thermal history, effect defined as color hysteresis (Kulčar et al. 2010, 2011; Ibrahim 2012).

Furthermore, diverse authors have also reported that the TC decolorization process is not complete, even at high temperatures. Through observation and color measurements, different results were attained across TC pigments' brands. Kulčar et al. (2010) observed a yellow shade above the activation temperature of the pigments handled, differing from the substrate color used. The possible causes discussed rely on a potential incomplete transparency of the leuco dye compound, light scattering the microcapsules' shell and/or blue light absorption of the binder (Kulčar et al. 2012). Ibrahim (2012) and Cabral and Souto (2014) observed a residual color in the colorless state, according to each TC pigment color, which can also indicate an incomplete dissociation of the color former and developer in the microcapsules. The shades remained in TC colorless state can present implications in the color change effect, being an important parameter to take in consideration particularly when specific colors are to be designed.

TC leuco dyes lifespan depends mostly upon the external conditions at which the materials are exposed to, rather than the number of times they are activated. Color fastness properties refer to the resistance of colored substances to remain stable under specific conditions (Aspland 1998). TC materials usually present poor stability to UV light and reasonable wash and rub fastness properties. Besides, they also show propensity to degrade when exposed to high temperatures (commonly referred above 140 °C) and aggressive solvents (Friškovec et al. 2013; Ferrara and Bengisu 2014; Periyasamy and Viková 2019). Ibrahim (2012) have researched the possibility to improve TC leuco dyes light fastness through the use of UV absorbers, hindered amine light stabilizers and antioxidants. Non-phenolic UV absorbers attained enhanced light stability, although TC daylight exposure for long periods of time demonstrated to be still limited.

2.4 Thermochromic Textiles

Early TC textiles were developed in the late 1980s, of which are examples: Sway®, a multicolor fabric presented by Toray Industries Inc. where activation temperatures were defined for specific end-uses such as ski-wear (11–19 °C), women's clothing (13–22 °C) and window shades (24–32 °C); and Hypercolor, a clothing line manufactured by Generra Sportswear, which heat-sensitive t-shirts became well known (Rhodes 1991; Rijavec and Bračko 2007). However, these examples have shown short lifespan and limited information was published in regards to their development. It was at the beginning of twenty-first century that ready-made TC materials began to be more easily accessed and thus, more systematically explored and researched in textiles (Ferrara and Bengisu 2014).

Thermochromic microcapsules were not initially developed for textile applications and present a lack of affinity and water insolubility. Thus, TC dispersion or powder form are commonly applied to textile substrates as pigments combined with appropriate binders, through printing processes (Christie 2013; Chowdhury et al. 2014). Screen printing has been the most approached technique, presenting great versatility for the study and manufacturing of TC color effects and patterns, also allowing the combination with other printing materials. Furthermore, screen printing also represents lower material waste in comparison to other techniques, which is an important consideration for sustainable reasons and given the expense of TC colourants in relation to conventional ones.

In this regard, great advantages are foreseen with inkjet printing, but it is not a feasible process yet due to the TC microcapsule size and dispersion in the solution (Ibrahim 2012). Christie, Shah and Wardman have conducted an experimental study with inkjet printing on cotton by microfiltration of the TC pigments before printing, having still observed the tendency of the TC microcapsules to clog the print heads (Ledendal 2015).

Dyeing processes with TC materials have been limited so far due to their water insolubility. Addressing this issue, a dyeing method was patented in 1992, comprising of a process to treat textile cellulose fibers with a cationic compound, followed by the immersion in a dispersion with TC and high polymer compound and further treatment with a binder addition (Masayasu et al. 1992), whereas no applications with this technique have been reported. More recently, Matsui Chromicolor® developed the 'Aqualite Coloring System', enabling a water-based TC paste with clear dilutant and thinners to be dyed on cotton (Matsui n.d.). This new TC material opens up emergent possibilities to explore and develop color change effects through dyeing processes.

So far, few researchers have reported the incorporation of TC materials directly into fibers production. Rubacha (2007) developed a process of TC cellulose fibers based on wet spinning from concentrated solvents of cellulose with TC leuco dyes. The developed fibers presented a chromic effect, but their physical and mechanical properties were low, which the author suggests that smaller TC microcapsules and narrower range of size distribution would improve them. Ibrahim (2012) has incorporated TC in powder form into different polymers by screw extrusion. He created color change filaments of polypropylene (PP), ethylene vinyl acetate (EVA) and linear low-density polyethylene (LLDPE), which also attained low mechanical properties.

The study of TC materials integration processes is an important subject to explore color change effects and through printing, diverse techniques have been addressed. The interrelation of chromic behavior with how pigments are combined and printed, as well as how thermal variation is introduced, has been catching special attention of researchers and designers with the objective to explore the potential of these materials to design interactive surfaces.

TC and conventional pigments can be combined in the same paste or overlapped individually during the printing process. In both cases, the combination possibilities are limitless considering the pigments' number, type and colors mixed or overlapped, concentrations and activation temperature of the TC pigments handled. While some of these parameters are associated to familiar processes when working with conventional colourants, the temporal and interactive dimensions of TC pigments changes how one is used to work with subtractive color mixtures.

Linda Worbin conducted a practice-based design research focused on dynamic textile patterns, with emphasis on chromic textiles. Through a series of design examples, she explored new dynamic qualities that these smart materials introduce to textiles and textile design (Worbin 2010). In Graffiti Cloth project, Worbin studied dynamic chromic palettes with TC and conventional pigments, applied through overprinting and raster techniques. The set of screen printed samples created a 'color map' of chromic possibilities with the pigments and raster scales combined. The researcher described the mixing process with TC pigments as more laborious and slower than with conventional pigments, due to the involvement of a heating phase to analyze colors in TC colorless state, nuances between transitions and the definition of possible adjustments. Furthermore, the process was also compared to mixing for three colors, the color of the TC pigment, conventional pigment and pigments overlapping.

The use of raster techniques allows the variation of color strength without changing the pigments' concentration in the paste and it is also applied for the halftone overprinting with CMYK separation. Celine Marcq explored this technique with TC pigments, to create an image reproduction with dynamic multicolors. The designer also incorporated a dichroic smart film in the substrate backside, that changes color when observed by different angles or moving the textile, thus adding a new variable in the creation of color change effects (Marcq 2014).

Furthermore, TC pigments have also been studied in combination with other chromic materials. Sara Robertson research included studies of color change effects with textiles screen printed with TC leuco dyes and coated with TC liquid crystals. The use of both TC systems, with different activation and threshold temperatures, lead to a broad chromatic palette and color change transitions through temperature variation (Robertson 2011). In the project Aqua chameleon swimming suit developed by Yun Ding, thermo, photo and hydrochromic pigments were applied through overlapping of different geometric and floral printing patterns, which enabled the creation of temporary expressions, according to the stimulus exposed: temperature, UV light and water (Collet 2007).

Zane Berzina investigated the use of TC leuco dyes with other microencapsulated materials. Focused on the human skin as a metaphor to develop sensing and reacting surfaces, the author combined TC pigments, aroma microcapsules and phase change materials in a wallpaper that when touched displayed: visual response by color change, olfactory response by releasing aroma and thermal variation according to the stored heat (Berzina 2004).

Geng et al. (2018) have designed and fabricated TC microencapsulated phase change materials with reversible property and thermal energy storage/release performance. Furthermore, they also researched the modification of these microcapsules with silver nano-particles, which also displayed antibacterial activity.

Textile colour and pattern change can encompasse diverse relations between forms composed by the printed pattern and the behaviour of the pigments and colours applied. In *Rather boring*, Worbin designed a textile pattern composed by aligned small forms printed with TC and conventional pigments of a similar grey colour. When the textile is heated, TC forms fade away and a message is revealed corresponding to the conventional pigments' pattern. The researcher highlights that the two pigments prints require an alignment with high precision, so before the textile is heated the regular pattern composition does not present differences, also alluding for the challenge to achieve colour similarities for each pigment type (Worbin 2010). The difficulty to reproduce specific colors with TC pigments is also discussed by Kulčar et al. (2012), referring it as an advantage for anti-counterfeit applications.

In addition to the TC pigments combination and printing possibilities, the heat generation source is also an important consideration during TC textiles design, as an intrinsic parameter of the chromic behavior and in respect to the relations between printed and heated patterns. According to the way the heating source(s) activates the changes, Worbin (2010) distinguishes chromic dynamic patterns as direct or reported. Direct patterns account for color change as a direct response to the thermal stimulus. Reported patterns involve the use of conductive and electronic components, which are programmed to induce temperature variation through resistive heating of conductive materials.

In direct patterns, transitions between TC pigments colorized and colorless states occur, in some degree, independently of pre-defined decisions of the designer, for example in which areas and when the textile is heated. In *Do Pattern* (Worbin 2010), cups with different three-dimensional shapes at the base were designed so that when filled with hot liquids interact with a tablecloth screen printed with TC and conventional pigments. The dynamic pattern depends on the action of the user, being co-author of the textile expressions by the way the cups are (re)placed on top of the TC textile and for how long they are kept at the same place. Other examples explore visual gestural or body print experiences, such as Swamp Stool by NunoErin and Bed linens by Jürgen Mayer (Quinn 2010); monitoring of body or breath temperature such as the thermochromic flu Mask by Kooroshnia (2011); and interaction with ambient temperature, which Marie Ledendal investigated through TC textiles exposition to sunlight (Ledendal 2015).

In reported patterns, sequences of activation are programmed to perform thermal variation of the conductive materials, that can be triggered through a specific stimulus or continuously run over time. Maggie Orth conducted early investigations in electronic textiles and explored the expressive potential of chromic displays through TC pigments screen printed in fabrics, woven with conductive threads (McQuaid 2005). In Orths' project Blip, the textile changes from monochromatic black tones in the cold state to bright colors in the heated state when programmed to occur in different areas at different periods of time (Orth 2010). Joanna Berzowska also used conductive threads to explore relations between the TC pigments pattern and the physical configuration of the electric circuit, woven in a Jacquard loom. Her interest in textiles dynamic behavior relied on the potential of soft responsive computation for "personal expression and playful exploration" (Berzowska 2004, 2005).

Furthermore, direct and reported patterns can also be addressed simultaneously. Kärt Ojavee and Eszter Ozsvald created SymbiosisS, a part of a collection of TC felt interfaces that when touched, changes color in response to the temperature of the human body and activates embedded electronics with tactile capacitive sensing, triggering gradual changes of a predefined conductive pattern (Ojavee and Ozsvald 2012; Ferrara and Bengisu 2014).

The stimulus that triggers the activation of color change can also assign different purposes, where textile expressions (color and pattern) can be linked to specific information. For example in Fabrication Bag project by Landin and Worbin (2005), the color of a bag dot pattern changes from dull to colorful, for a visual and expressive alert that the bag user received a text message or a call; in Warning Signs, by Ngo and Lam (2010), a sweater pattern is revealed through color change, to communicate the presence of high levels of carbon monoxide.

The selection of the conductive materials' type, integration processes in the textile substrate and activation definition has a significant effect on the dynamic textile behavior and expressions. This topic will be further discussed in Chap. 4.

Through diverse conceptual and technical perspectives, the presented projects discuss TC textiles simultaneously as a physical structure and a system of interactions. Inherent dynamic possibilities and development processes, lead to important findings and knowledge in order to design with chromic materials and textiles. From the understanding of these materials behavior and interrelations with printing techniques, dynamic color effects and heating stimulus, less emphasis has been placed on how to combine TC and conventional pigments through systematic processes. This topic is omitted in some studies or referred to as an empirical process involving trial and error paste combinations with direct observation and comparison between samples. The experimental work developed in this research encompassed systematic processes. Other approaches such as in the Graffiti Cloth project (Worbin 2010), developed a collection of color swatches that through defined characteristics of the pigments, colors, concentrations and printing methods applied, build up a visual tool for color change analysis, chromatics selection and elaboration techniques.

Furthermore, the intrinsic behavior of TC pigments in response to heat entails complex relations between TC pigments activation temperature(s) selected, solutions to induce thermal variation, where and when the changes occur, which expressions and the stimulus that triggers it (reported patterns). These are important parameters to consider in the design of chromic textile behavior and to explore new possibilities with this dynamic and interactive material.

References

Abel A (2012) The history of dyes and pigments: from natural dyes to high performance pigments. In: Best J (ed) Colour design: theories and applications. Woodhead Publishing and Textile Institute, Oxford

Addington M, Schodek D (2005) Smart materials and new technologies. Architectural Press, Amsterdam

Aitken D, Burkinshaw SM, Griffiths J, Towns AD (1996) Textile applications of thermochromic systems. Rev Prog Color Relat Top 26:1–8. https://doi.org/10.1111/j.1478-4408.1996.tb00105.x

Albers J (1968) Oral history interview with Josef Albers, 1968 June 22–July 5 [Online]. https://www.aaa.si.edu/collections/interviews/oral-history-interview-josef-albers-11847: Archives of American Art, Smithsonian Institution. Accessed 13 March 2012

Albers J (2013) Interaction of color. Yale University Press, New Haven

Aspland JR (1998) Colorants: dyes. In: Nassau K (ed) Color for science, art and technology. Elsevier, Amsterdam

Bamfield P, Hutchings MG (2010) Chromic phenomena: technological applications of colour chemistry. Royal Society of Chemistry, Cambridge

Baurley S (2004) Interactive and experiential design in smart textile products and applications. Pers Ubiquit Comput 8:274–281. https://doi.org/10.1007/s00779-004-0288-5

Berzina Z (2004) Skin stories: charting and mapping the skin. Research using analogies of human skin tissue in relation to my textile practice. PhD thesis, University of the Arts London

Berzowska J (2004) Very slowly animating textiles: shimmering flower. ACM SIGGRAPH 2004 sketches. ACM, Los Angeles, California

Berzowska J (2005) Electronic textiles: wearable computers, reactive fashion, and soft computation. Textile 3:58–75. https://doi.org/10.2752/147597505778052639

Bloj M, Hedrich M (2016) Color perception. In: Chen J, Cranton W, Fihn M (eds) Handbook of visual display technology, 2nd edn. Springer, Berlin

Boyce PR (2014) Human factors in lighting, 3rd edn. Taylor & Francis, London

Brainard D (2003) Color appearance and color difference specification. In: Shevell SK (ed) The science of color. Elsevier, Amsterdam

Briggs-Goode A (2013) Printed textile design. Laurence King Publishing, London

Brownell B (2017) Transmaterial next: a catalog of materials that will redefine our future. Princeton Architectural Press, New York

Büttikger U (1993) Light and space. Birkhäuser, Basel

Cabral I, Souto AP (2014) Thermochromic textile structures: a dynamic ambient light design. Int J Des Objects 7(3):23–35. https://doi.org/10.18848/2325-1379/CGP/v07i03/38683
Cassidy T (2018) Color knowledge. In: Cassidy T, Goswami P (eds) Textile and clothing design technology. CRC Press, Boca Raton
Chilton J, Lau B (2015) Lighting and the visual environment in architectural fabric structures. In: Llorens J (ed) Fabric structures in architecture. Woodhead Publishing, Cambridge
Choudhury A (2010) Scales for communicating colours. In: Gulrajani ML (ed) Colour measurement: principles, advances and industrial applications. Woodhead Publishing, Cambridge
Choudhury AKR (2014a) Principles of colour and appearance measurement: visual measurement of colour, colour comparison and management, vol 2. Woodhead Publishing, Cambridge
Choudhury AKR (2014b) Principles of colour appearance and measurement Volume 1: Object appearance, colour perception and instrumental measurement. Woodhead Publishing, Cambridge
Chowdhury MA, Butola BS, Joshi M (2013) Application of thermochromic colorants on textiles: temperature dependence of colorimetric properties. Color Technol 129:232–237. https://doi.org/10.1111/cote.12015
Chowdhury MA, Joshi M, Butola BS (2014) Photochromic and thermochromic colorants in textile applications. J Eng Fabr Fibers 9(1):107–123
Christie RM (2013) Chromic materials for technical textile applications. In: Gulrajani ML (ed) Advances in the dyeing and finishing of technical textiles. Woodhead Publishing, Oxford
Christie RM, Robertson S, Taylor S (2007) Design concepts for a temperature-sensitive environment using thermochromic colour change. Colour Des Creat 1(1):5, 1–11
Clarke S (2011) Textile design. Laurence King, London
Clarkson J (2008) Human capability and product design. In: Schifferstein H, Hekkert P (eds) Product experience. Elsevier, San Diego, CA
Collet C (2007) The next textile revolution. In: Bonnemaison S, Macy C (eds) Responsive textile environments. TUNS Press, Halifax, N.S.
Cruz P (2016) Structures and architecture: beyond their limits. CRC Press, London
Cuttle C (2015) Lighting design: a perception-based approach. Routledge, London and New York
Das T (2009) Surface design of fabrics for interior textiles. In: Rowe T (ed) Interior textiles: design and developments. CRC Press, Boca Raton
Day JH (1963) Thermochromism. Chem Rev 63:65–80. https://doi.org/10.1021/cr60221a005
Descottes H, Ramos CE (2011) Architectural lighting: designing with light and space. Princeton Architectural Press, New York
Dickinson K (2011) The use of colour in textile design. In: Briggs-Goode A, Townsend K (eds) Textile design: principles, advances and applications. Woodhead Publishing and The Textile Institute, Oxford
Du D (2011) Interior lighting. Profession Design Press, California
Durasevic V (2016) Smart dyes for medical textiles. In: Langenhove LV (ed) Advances in smart medical textiles. Woodhead Publishing, Oxford
Eastaugh N, Walsh V, Chaplin T, Siddall R (2004) The pigment compendium: a dictionary of historical pigments. Elsevier Butterworth-Heinemann, Oxford
Eliasson O (n.d.) Your atmospheric colour atlas, 2009 [Online]. http://olafureliasson.net/archive/artwork/WEK100285/your-atmospheric-colour-atlas. Accessed 8 Feb 2017
Feisner EA (2006) Color studies. Fairchild Publications, New York
Ferrara M, Bengisu M (2014) Materials that change color: smart materials, intelligent design. Springer International Publishing, Cham
Field GG (1998) Color imaging: printing and photography. In: Nassau K (ed) Color for science, art and technology. Elsevier, Amsterdam
Friškovec M, Kulčar R, Gunde MK (2013) Light fastness and high-temperature stability of thermochromic printing inks. Color Technol 129:214–222. https://doi.org/10.1111/cote.12020
Gangakhedar N (2010) Colour measurement methods for textiles. In: Gulrajani ML (ed) Colour measurement: principles, advances and industrial applications. Woodhead Publishing, Cambridge
Ganslandt RD, Hofmann H (1992) Handbook of lighting design. ERCO, Braunschweig/Wiesbaden

Geng X, Li W, Yin Q, Wang Y, Han N, Wang N, Bian J, Wang J, Zhang X (2018) Design and fabrication of reversible thermochromic microencapsulated phase change materials for thermal energy storage and its antibacterial activity. Energy 159:857–869. https://doi.org/10.1016/j.energy.2018.06.218

Gerber A (n.d.) King Fahad National Library [Online]. https://www.gerberarchitekten.de/en/project/king-fahad-national-library/. Accessed 10 Feb 2017

Goethe J, Eastlake C (2010) Goethe's theory of colours. LaVergne, TN, Kessinger Publishing

Goldstein BE (2010) Sensation and perception. Wadsworth, Cengage Learning, Belmont, CA

Goodman TM (2012) International standards for colour. In: Best J (ed) Colour design: theories and applications. Woodhead Publishing and Textile Institute, Oxford

Gregory RL (1978) Eye and brain: the psychology of seeing. McGraw-Hill, New York

Gupte V (2010) Expressing colours numerically. In: Gulrajani ML (ed) Colour measurement: principles, advances and industrial applications. Woodhead Publishing, Cambridge

Hanson AR (2012) What is colour? In: Best J (ed) Colour design: theories and applications. Woodhead Publishing and Textile Institute, Oxford

Hidefi M (2012) Understanding and forecasting colour trends in design. In: Best J (ed) Colour design: theories and applications. Woodhead Publishing and Textile Institute, Oxford

Hunt RWG (2004) The reproduction of colour. Wiley, Chichester

Hurlbert A, Ling Y (2012) Understanding colour perception and preference. In: Best J (ed) Colour design: theories and applications. Woodhead Publishing and Textile Institute, Oxford

Ibrahim W (2012) An investigation into textile applications of thermochromic pigments. PhD thesis, Heriot-Watt University

Ingamells W (1993) Colour for textiles: a user's handbook. Society of Dyers and Colourists, Bradford

Innes M (2012) Lighting for interior design. Laurence King Publishing, London

Itten J (1973) The art of color: the subjective and objective rationale of color. Van Nostrand Reinhold Co., New York

Itten J, Birren F (1970) ITTEN the elements of color. Van Nostrand Reinhold Co., New York

Kalat JW (2007) Biological psychology. Thomson/Wadsworth, Belmont, CA

Karlen M, Benya J, Spangler C (2012) Lighting design basics. Hoboken, NJ, John Wiley

Kooroshnia M (2011) Mask [Online]. http://hb.diva-portal.org/smash/record.jsfpid=diva2%3A884770&dswid=-1987-sthash.ymQP0NI7.dpbs: University of Borås. Accessed 19 March 2017

Krauskopf J (1998) Color vision. In: Nassau K (ed) Color for science, art and technology. Elsevier, Amsterdam

Kries M, Kugler J (2013) Lightopia. Vitra Design Museum, Weil am Rhein

Kuehni RG (2012) Color: an introduction to practice and principles. Wiley, Hoboken

Kulčar R, Friškovec M, Gunde MK, Knešaurek N (2011) Dynamic colorimetric properties of mixed thermochromic printing inks. Color Technol 127:411–417. https://doi.org/10.1111/j.1478-4408.2011.00338.x

Kulčar R, Friškovec M, Hauptman N, Vesel A, Gunde MK (2010) Colorimetric properties of reversible thermochromic printing inks. Dyes Pigm 86(3):271–277. https://doi.org/10.1016/j.dyepig.2010.01.014

Kulčar R, Gunde MK, Nina Knešaurek N (2012) Dynamic colour possibilities and functional properties of thermochromic printing inks. acta graphica 23:25–36

Lam W (1992) Perception and lighting as formgivers for architecture. Van Nostrand Reinhold, New York

Landin H, Worbin L (2005) The fabrication bag—an accessory to a mobile phone. Ambience 05, 2005 Tampere, Finland

Lechner N (2015) Heating, cooling, lighting: sustainable design methods for architects. Wiley, New Jersey

Ledendal M (2015) Thermochromic textiles and sunlight activating systems: an alternative means to induce colour change. PhD thesis, Heriot-Watt University

Lennie P (2003) The physiology of color vision. In: Shevell SK (ed) The science of color. Elsevier, Amsterdam
Maack T, Pawlik K (2009) Light perspectives between culture and technology. Erco, Lüdenscheid
Mahapatra N (2016) Textile dyes and dyeing. Woodhead Publishing India, New Delhi
Marcq C (2014) Dynamic prints [Online]. http://etextile-summercamp.org/2015/dynamic-prints/. Accessed 27 Sept 2015
Marcus RT (1998) The measurement of colour. In: Nassau K (ed) Color for science, art and technology. Elsevier, Amsterdam
Masayasu K, Osamu S, Shouzou S, Tatsuya M (1992) Dyeing method and dyed product
May M (2007) Sensation and perception. Chelsea House Publishers, New York
Mcauliffe M (2016) The perception of light. Oxford Global Press, London
Mcquaid M (2005) Extreme textiles: designing for high performance. Smithsonian Cooper-Hewitt, National Design Museum, New York
Nassau K (1998) Fundamentals of colour science. In: Nassau K (ed) Color for science, art and technology. Elsevier, Amsterdam
Ngo S, Lam N (2010) Warning signs [Online]. http://www.itp.nyu.edu/~sn964/SueITPblog/wearables-finalwarning-signs/. Accessed 30 April 2013
Nilsson L, Satomi M, Vallgårda A, Worbin L (2011) Understanding the complexity of designing dynamic textile patterns. In: Proceedings of the Ambience conference, 2011 Borås, Sweden
Ojavee K, Ozsvald E (2012) SymbiosisS. Leonardo 45:388–389. https://doi.org/10.1162/LEON_a_00422
Orth M (2010) Blip, 2010 [Online]. http://www.maggieorth.com/art_Blip.html. Accessed 23 Jan 2013
Parraman C (2012) Colour printing techniques. In: Best J (ed) Colour design: theories and applications. Woodhead Publishing and Textile Institute, Oxford
Pérez V, Saiz D, Verdú F (2010) Colour vision: theories and principles. In: Gulrajani ML (ed) Colour measurement: principles, advances and industrial applications. Woodhead Publishing, Cambridge
Periyasamy AP, Viková M (2019) Testing of chromic materials. In: Viková M (ed) Chromic materials: fundamentals, measurements, and applications. Apple Academic Press, Waretown
Quinn B (2010) Textile futures: fashion, design and technology. Berg, Oxford
Rhodes E (1991) To dye for hypercolor clothes heating up market for maker Generra [Online]. http://community.seattletimes.nwsource.com/archive: The Seattle Times. Accessed 3 Feb 2013
Rijavec T, Bračko S (2007) Smart dyes for medical and other textiles. In: Langenhove LV (ed) Smart textiles for medicine and healthcare: materials, systems and applications. Woodhead Publishing, Cambridge
Rizzi A, Bonanomi C (2012) Colour illusions and the human visual system. In: Best J (ed) Colour design: theories and applications. Woodhead Publishing and Textile Institute, Oxford
Robertson S (2011) An investigation of the design potential of thermochromic textiles used with electronic heat-profiling circuitry. PhD thesis, Heriot-Watt University
Rubacha M (2007) Thermochromic cellulose fibers. Polym Adv Technol 18:323–328. https://doi.org/10.1002/pat.889
Salaün F (2016) Microencapsulation technology for smart textile coatings. In: Hu J (ed) Active coatings for smart textiles. Woodhead Publishing, Oxford
Schwartz MM (2002) Encyclopedia of smart materials. Wiley, New York
Sear F (2008) Roman architecture. Routledge, London
Seeboth A, Lötzsch D (2008) Thermochromic phenomena in polymers. Smithers Rapra, Shrewsbury
Seeboth A, Lötzsch D (2013) Thermochromic and thermotropic materials. Pan Stanford, Singapore
Setchell JS (2012) Colour description and communication. In: Best J (ed) Colour design: theories and applications. Woodhead Publishing and Textile Institute, Oxford
Shao J, Liu G, Zhou L (2016) Biomimetic nanocoatings for structural coloration of textiles. In: Hu J (ed) Active coatings for smart textiles. Woodhead Publishing, Oxford
Sherin A (2012) Design elements color fundamentals. Rockport Publishers, Beverly, MA

Smith V, Pokorny J (2003) *Color matching and color discrimination*. In: Shevell SK (ed) The science of color. Elsevier, Amsterdam

Talvenma P (2006) Introduction to chromic materials. In: Mattila H (ed) Intelligent textiles and clothing. Woodhead, Cambridge

Tilley RJD (2000) Colour and optical properties of materials: an exploration of the relationship between light, the optical properties of materials and colour, Wiley, Chichester

Triedman K (2015) Colour: the professional's guide: Understanding, appreciating and mastering colour in art and design. Ilex, Lewes

Turrell J (n.d.) Two blues (2008) [Online]. http://jamesturrell.com/work/two-blues/. Accessed 8 Feb 2017

Vik M (2019) Basic terms. In: Viková M (ed) Chromic materials: fundamentals, measurements, and applications. Apple Academic Press, Waretown

Viková M (2019) Type of chromic materials. In: Viková M (ed) Chromic materials: fundamentals, measurements, and applications. Apple Academic Press, Waretown

Wardman RH (2018) An introduction to textile coloration: principles and practice. Wiley, Hoboken

Westland S (2016) Color communication. In: Chen J, Cranton W, Fihn M (eds) Handbook of visual display technology, 2nd edn. Springer, Berlin

Worbin L (2010) Designing dynamic textile patterns. Doctoral thesis, University of Borås

Xin JH (2006) Total colour management in textiles. Woodhead Publishing and The Textile Institute, Cambridge

Yavuz G, Zille A, Seventekin N, Souto AP (2018) Structural coloration of chitosan coated cellulose fabrics by electrostatic self-assembled poly(styrene-methyl methacrylate-acrylic acid) photonic crystals. Carbohydr Polym 193:343–352. https://doi.org/10.1016/j.carbpol.2018.03.084

Yot R (2011) Light for visual artists: understanding & using light in art & design. Laurence King, London

Yuan W, Wu C, Zhou N, Zhang K (2015) Fibers with the tunable structure colors based on the ordered and amorphous structures. In: Tao X (ed) Handbook of smart textiles. Springer, Singapore

Chapter 3
Experimental Work: Thermochromic Leuco Dyes

Development of paste recipes with TC and conventional pigments for textile printing is commonly realized through empirical processes. Towards a systematic and more accurate process to create TC textiles, this chapter presents experimental work based on instrumental processes of paste recipe formulation and respective integration in textile substrates.

Contextualizing the objective to change light transmittance through the textile chromic behavior, a preliminary analysis was conducted on the interaction between light and textile dynamic color. After, the research focused on the development of screen printed textiles with defined color change ratios: from different to similar colors upon heating, as well as the inverse.

3.1 Introduction

Thermochromic leuco dyes exhibit a reversible change from colored to colorless when exposed above their activation temperature. By combining TC with conventional pigments, the chromic effect attains a change from one color to another (Worbin 2010; Viková 2019).

In this chapter, TC pigments were researched towards the development of TC textiles, which dynamic behavior presents predefined color ratios through temperature variation. The design concept inherent to the materials' research hinges on the interaction of the textile chromic behavior and light, aiming to transform similar light intensities and tones to heterogeneous, as well as the inverse.

Dark colors absorb a vaster intensity of the visible light spectrum than lighter colors (Descottes and Ramos 2011). Thus, TC textiles change the light that passes through them differently, when they are below or above their activation temperature. A preliminary analysis was conducted in order to study light transmittance variation created by the chromic behavior of textiles screen printed with different TC pigments' colors, concentrations and combination processes with conventional colourants.

I. D. Cabral et al., *Dynamic Light Filters*, Springer Series in Design and Innovation 3, https://doi.org/10.1007/978-3-030-39529-2_3

The experimental work concerned with the development of TC textiles that change color according to predefined ratios has involved two main studies, focused on processes of paste recipe elaboration with TC and conventional pigment and respective application in textile substrates. First, the objective was to study the combination of pigments' colors and concentrations towards screen printed textiles that change from different to similar colors with temperature increase. After, the colourimetric properties of the pigments handled were measured and the results applied in the development of a database for paste recipes formulations, which colors change from similar to different upon the thermal stimulus. Besides screen printing processes, this study also included the development of a study prototype through rotary printing (Sect. 3.4).

Figure 3.1 describes the materials, equipment and methods applied during the experimental work discussed in this chapter. Textile substrate 1 was applied in the experimental work of Sects. 3.2, 3.3 and 3.4; and textile substrate 2 was solely used for the study and development of a prototype presented in Sect. 3.4. All procedures were conducted at a room temperature below the activation temperature of the TC pigments used (27 °C), except when specified otherwise.

3.2 Textile Color Change and Light Transmittance

For a preliminary analysis of the light transmittance variation that color change textiles may produce, two studies were conducted. First, the relation of light and color change was studied through screen printed samples with different TC pigment colors and concentrations in the paste. After, the phenomenon was analyzed through samples, which application of TC and conventional colourants in the textile substrate was developed by different processes.

To study how color change may transform the incident light that passes through TC textiles, a set of samples was developed comprising of five samples of each TC pigment color: orange, red, magenta, blue and black, at 10% concentration (samples A to E, respectively); three samples with TC pigment black at different concentrations in the paste: 10; 5 and 1% (samples E to G, respectively); and three samples, which combination of TC and conventional colourants was applied to the textile substrate by different processes: screen printing of TC and conventional pigments combined in the same paste (H), overprinting of conventional and TC pigments in different pastes (I) and screen printing of TC pigment in a pre-dyed substrate with conventional colourants (J). Table 3.1 presents the samples' description.

Chromic behavior effect on light transmittance was analyzed by direct observation, photographic record and light intensity measurement. The study was conducted in a dark room, where each sample was placed on the open face of a light box. In the photo record, the light box was set with the open face perpendicular to a white wall and a camera set on a tripod. Camera settings were equal for all photos recorded, for a comparison between luminosities. In the light intensity measurement, the open face of the light box was set in vertical position, 1 m from the luxmeter in the first

MATERIALS

Textile substrate 1: 50% cotton (CO) 50% polyester (PES) plain weave, 115g/m2
15,0 warp and 15,5 weft (tex), density of 41 ends/cm and 31 picks/cm

Textile substrate 2: 100% CO 3/1 twill weave, 190g/m2
25,0 tex, warp and weft yarns, density of 46 ends/cm and 25 picks/cm

Pigments: TC pigments Varioterm AC base 27 (27ºC heat sensitive pigment) - orange, red, magenta, blue and black

Conventional pigments ATUSMIC Magnaprint eco yellow HR, orange HD, red HB, pink H5B, Blue HB and Black H3B

Binder: Screen printing pastes were formulated with vinyl acrylic binders, having being used two brands: Atusmic Magnaprint ND (samples in section 3.4) and Gilaba (samples in sections 3.3 and 3.5)

Dyes: Disperse dye Dianix orange SG 200%

Reactive dye Levafix orange CA

EQUIPMENT & METHODS

Screen printing processes were conducted in a Zimmer Mini MDF R541 table, with the following settings: magnetic field level 3, application speed 50% of the graded dial and higher rod diameter parameter. Pastes were applied with a metallic rod-squeegee of 12 mm, using an open screen with 101,6 TPI and completed a process of drying and thermo setting in a Werner Mathis AG laboratory oven at 150ºC for 3 minutes.

Rotary printing processes were developed in an industrial context to manufacture of a study prototype (section 3.4). Four cylinders were used, having each 150 cm width and 91,4 cm diameter with a 125 mesh (openings per linear inch).

Dyeing processes were conducted in an Ibelus C720 equipment. For the PES fibres of the textile substrate 1, a Dianix orange SG 200% disperse dye was used at 1% concentration in a bath ratio of 1/10 and with pH 4,5. CO fibres were dyed with Levafix orange CA gran reactive dye at 1% concentration, in a bath ratio of 1/10 with 35g/L of salt and 10g/L of sodium carbonate. Dyeing programs applied for disperse and reactive dyes are presented in the graphics left and right, respectively.

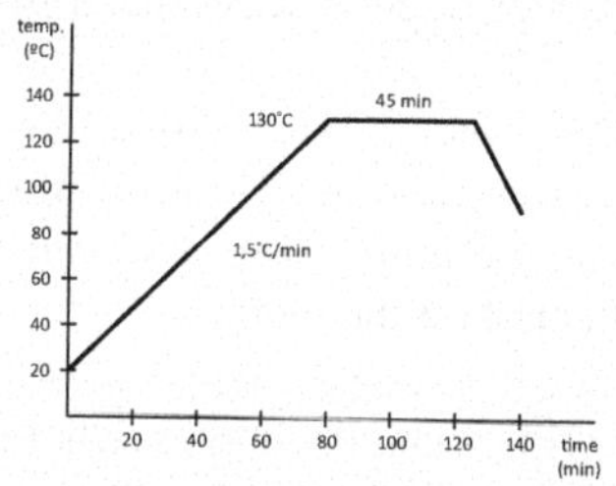

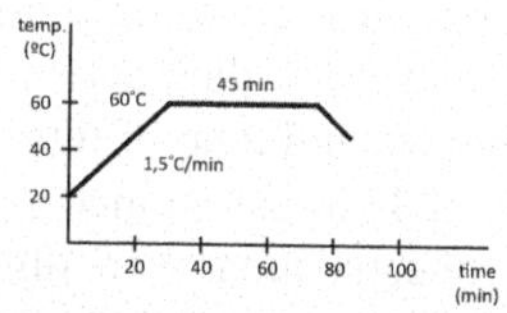

To analyse the light transmittance variation, a light box with a 28 x 18 cm aperture was used to place the textile samples and the light intensity was measured with a DeltaOhm – HD 2302.0 luxmeter. The experiments were conducted with different setups as will be described in the experimental sections.

A spectrophotometer Datacolor International, SF600 Plus – CT with Datacolor Match 3.4 software was used to conduct spectral reflectance measurements, data analysis and paste recipes formulation. This equipment measures the amount of light reflected at each wavelength of visible light spectrum (approximately 380 to 760 nm), according to the average color on a defined aperture (Xin, 2006). Previous to the measurements, the spectrophotometer was calibrated with a black trap and white tile. The measurements were developed with the illuminant D65, LAV/Specular Included, standard observer 10º and an aperture measurement of 9 mm. Each sample was measured in three points.

Fig. 3.1 Materials, equipment and methods applied in the experimental work

Table 3.1 Description of samples A–J

Sample	Description
A	Screen printed with 10% TC orange
B	Screen printed with 10% TC red
C	Screen printed with 10% TC magenta
D	Screen printed with 10% TC blue
E	Screen printed with 10% TC black
F	Screen printed with 5% TC black
G	Screen printed with 1% TC black
H	Screen printed with a mixture of 10% TC black and 1% conventional pigment orange
I	Screen printed with 1% conventional pigment orange, dried and screen printed with 10% TC black
J	Dyed with 1% orange disperse dye and 1% orange reactive dye, dried and screen printed with 10% TC black

measurement cycle and 0.5 m in the second. After the lamp was turned on, luxmeter values were recorded on three points of each sample. A dryer was used to heat up the TC samples above 27 °C.

Additionally, for a comparative analysis of the TC samples in decolorized state, a spectral reflectance measurement was conducted in the spectrophotometer and data was set in color coordinates and CIELAB differences. To maintain the samples heated during color measurement, a hot water rubber deposit was placed in the samples' backside.

Figure 3.2 presents the photos of samples screen printed with different TC colors, placed on the open face of the light box. Differences of light transmittance between samples, below and above their activation temperature, were observed and compared through the blur size and tone of light reflected on the wall.

Below 27 °C, light transmittance tones were perceived accordingly to each sample color and light reflected areas were higher with A and B samples, whereas with D and E, they were slightly perceived. Above 27 °C, similar luminosities were expected, as the TC pigments were in the decolorized state and all samples were developed

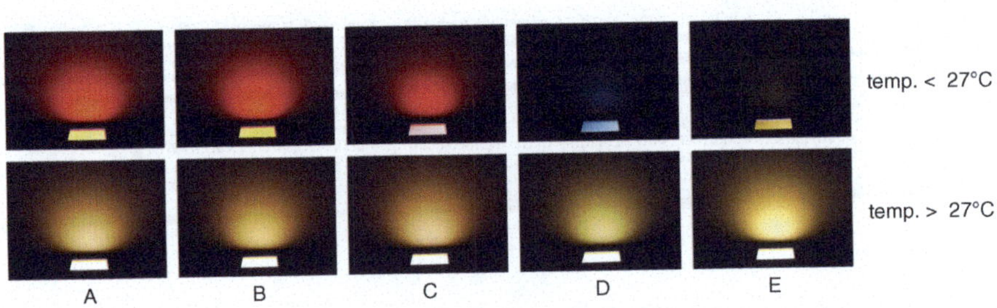

Fig. 3.2 Light transmittance variation with samples A, B, C, D and E

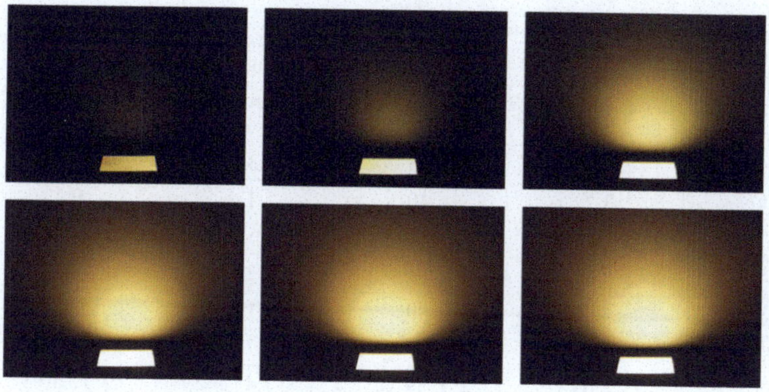

Fig. 3.3 Light transmittance variation with samples E, F and G

with the same textile substrate. However, small differences were observed: while A and B samples presented similar areas of light reflected on the wall and light tone, C appeared to be slightly larger and more red, D was slightly smaller and less red and E presented the highest light intensity and was more yellow than the other samples. Comparing light transmittance variation with temperature increase, it was observed that A and B samples presented the lower differences and the highest was attained with E sample.

Figure 3.3 presents the photos of screen printed samples with different concentrations of the TC pigment black: 10, 5 and 1% from E to G, respectively. Along the decrease of the pigment concentration below 27 °C, light transmittance increases and thus, light variations decrease with color change. Above 27 °C, sample G appeared to have a slightly higher light projection area in relation to E and F samples.

Table 3.2 presents the mean values of light intensities attained with samples A to G and respective percentual luminosity differences with temperature variation. At the cold state, light intensity values presented great differences between samples. A and B samples (TC orange and TC red) attained the higher light intensities of samples screen printed with 10% TC, with values approximately 400% higher than D and E (TC blue and TC black). With the concentration decrease of TC pigment black, light intensity increased approximately 100% from E to F (10–5%) and 500% from E to G (10–1%). At the heated state, sample D (TC blue) attained the lowest light intensity value and C (TC magenta) had the highest, with a difference of approximately 11%. With TC black samples, light transmittance of E was approximately 16% lower than G.

The percentual luminosity difference with temperature variation is higher for samples D and E. The pigment's concentration decreases from E to F samples represented a luminosity variation reduction of 12% and from F to G it was 38%. Moreover, while sample B (10% TC red) and G (1% TC black) attained different light intensities at each temperature state, the percentual luminosity difference was similar.

Table 3.2 Mean values of light intensities with samples A–G

Sample	Light intensity (lx)		Luminosity difference (%)	Light intensity (lx)		L di (%
	Temp. < 27 °C	Temp. > 27 °C		Temp. < 27 °C	Temp. > 27 °C	
A	2.13	3.84	45	5.32	9.54	4
B	2.23	3.72	40	5.43	9.43	4
C	1.42	4.07	65	3.52	10.24	6
D	0.47	3.61	87	1.22	9.40	8
E	0.45	4.00	89	1.19	10.57	8
F	0.95	4.20	77	2.72	11.53	7
G	2.82	4.63	39	7.45	12.30	3
	Distance sample/luxmeter = 1 m			Distance sample/luxmeter = 0.5 m		

The results attained demonstrate the dependency of the TC pigments' colors and concentrations applied, for the magnitude and ratio of light transmittance variation, in regards to light intensity and tone parameters. Furthermore, percentual luminosity differences in each sample for measurements conducted at 1 m and 0.5 m confirmed similarities. Further tests were carried at 1 m distance between textile sample and luxmeter.

In the photo record and light intensity measurement, TC samples presented differences in the colorless state, when similarities were expected, as previously mentioned. To analyze further these results, a color measurement was conducted with samples heated above 27 °C. Table 3.3 presents the results attained for each sample and the textile substrate applied. Data was set in color coordinates and CIELAB differences, where the substrate was set as the pattern color.

In decolorized state, all screen printed samples are darker than the substrate and E (10% TC black) presented the lower lightness value. TC samples were also more green and yellow than the substrate, except A and C which presented higher a* values and B had similar results. CIELAB differences varied between samples' colors, with sample C attaining the lower value (8.0) and E highest (13.2). According to the

Table 3.3 Color coordinates and CIELAB differences of samples A–G, above 27 °C

Sample	L*	a*	b*	C*	h	dE*
Substrate	93.93	2.77	−7.26	7.77	290.92	–
A	89.59	8.57	2.56	8.94	16.62	12.2
B	92.06	2.66	4.30	5.05	58.29	11.7
C	87.24	6.15	−4.35	7.54	324.74	8.0
D	90.43	−3.40	1.76	3.83	152.69	11.5
E	85.94	−2.21	2.04	3.01	137.27	13.2
F	88.24	−1.10	0.13	1.11	173.49	10.1
G	91.48	0.88	−3.59	3.70	283.81	4.8

reduction of TC pigment concentration in the pastes, there was also a decrease of color difference: G is approximately 2.8 times lower than E. The results attained suggest an incomplete decolorization of the TC pigments handled, which varies differently in regards to the pigments' color and concentration in the paste (Cabral and Souto 2014).

In regards to samples developed by different application processes of TC and conventional colourants, Fig. 3.4 presents the photos of light transmittance variation recorded. Below the TC pigments' activation temperature, sample I attained a larger light projection on the wall than H and J. When heated, sample I also appeared to have a slightly higher blur size and the smallest was J, which was also more red. It was not clearly perceived which sample performed greater light transmittance variation.

Mean values of light intensities presented in Table 3.4, confirmed the observations previously described and showed that the percentual luminosity difference is equal for H and J samples while I was 5% lower. Although below 27 °C, light transmittance with sample I was lower than with H, by direct observation of samples before placement in the light box, I sample appeared darker. The overlapping layers with the TC black at the surface, created an apparent darker color in relation to sample H, where TC and conventional pigments were mixed in the same paste, but the light transmittance results indicated that less TC pigment was absorbed.

Table 3.5 presents the color measurement results obtained with samples H, I and J, below and above 27 °C. Color coordinates of orange screen printed and orange dyed

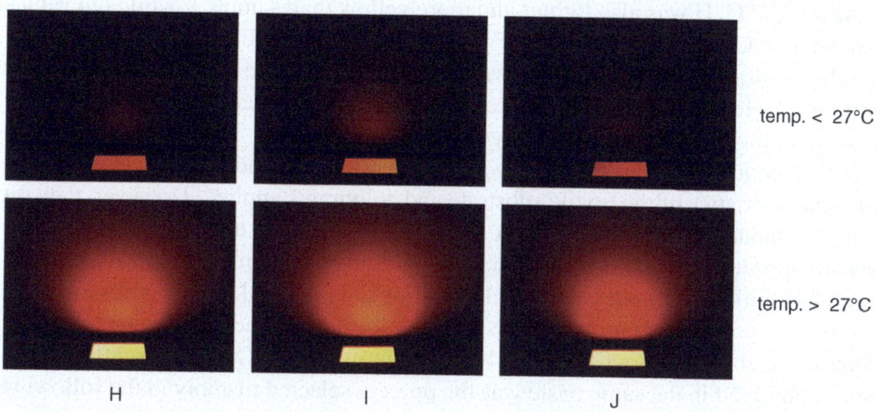

Fig. 3.4 Light transmittance variation with samples H, I and J

Table 3.4 Mean values of light intensities with samples H, I and J

Sample	Light intensity (lx)		Luminosity difference (%)
	Temp. < 27 °C	Temp. > 27 °C	
H	0.27	1.89	86
I	0.38	2.03	81
J	0.23	1.61	86

Table 3.5 Samples' color coordinates below and above 27 °C and CIELAB differences between heated samples and respective conventional colorant substrate

	Sample	L*	a*	b*	C*	h	dE*
	Orange print	61.14	51.88	54.17	75.01	46.24	–
	Orange dyed	61.14	42.31	43.49	60.68	45.79	–
Temp. < 27 °C	H	39.53	13.65	14.24	19.73	46.20	–
	I	30.46	10.43	7.94	13.11	37.26	–
	J	32.79	11.03	11.52	15.96	46.25	–
Temp. > 27 °C	H	60.34	44.16	45.66	63.52	45.96	11.5
	I	58.41	43.87	42.43	61.03	44.04	14.5
	J	58.34	38.81	43.93	58.62	48.54	4.5

substrates were also included for a comparison of the results with the samples heated. CIELAB differences respect to orange print as pattern color for H and I samples and orange dyed for J.

Below 27 °C, H and I attained the greater differences of the three samples, being H lighter, more red and yellow than I. According to the previous observations, the results demonstrate the effect of pigments' application processes in color appearance of the textiles, with emphasis in the lightness axis: L* value distance between H and I samples was approximately 9.

Above 27 °C, H was also lighter and more yellow than sample I, while red values were similar. Comparing the samples' colors above 27 °C with the respective pattern sample, the highest color difference was attained with sample I (14.5) and the lowest with J (4.5). Hence, the application processes of TC and conventional colourants also interfere in the TC pigment residual color.

In this section, the values attained with the experiments conducted relate to the specific characteristics of the textile substrate and colourants applied. However, they set up a preliminary understanding of the TC pigments handled and light transmittance variation produced by textile color change regarding TC pigment colors, concentrations and application processes, also delineating the potential to further research TC textiles with predefined chromic ratios, towards the development of dynamic lighting scenarios (Cabral and Souto 2012b). Screen printing with TC and conventional pigments combined in the same paste was the process selected to apply in the following experimental sections, with the objective to develop systematic processes of paste recipe elaboration.

3.3 Color Change: From Different to Similar Colors

In this section, the combination of TC and conventional pigments in screen printing pastes was researched towards the development of textiles that change from different to similar colors with temperature increase. The study conducted focused on colors and concentration ratio between TC and conventional pigments in respect to the color selection to maintain constant in samples' colorless state and colors palette in the cold state, as well as on the variation of TC pigment and binder concentrations in the color change effect.

Initially, a base framework of screen printed samples was developed, with the objective to build a visual tool of colors combination with the pigments handled, at defined concentrations. The screen printing pastes combine 5% concentration of one color of each pigment type, totalizing 10% pigments and 90% binder in each sample. Due to the color change objective, TC and conventional pigments with the same color were not screen printed.

After, a second framework was produced to analyze how the variation of the pigments' concentration affects samples' colors, below and above 27 °C. This study was conducted with mixtures of conventional pigment yellow and TC pigment orange, magenta, blue or black. All screen printing pastes have 90% binder, the conventional pigment percentage varies from 5; 4; 3; 2; 1–0.5% and the TC pigments complete the remaining percentages: 5; 6; 7; 8; 9 and 9.5%, respectively. Samples of both frameworks were analyzed and compared by direct observation, below and above 27 °C.

Additionally, the effect of TC pigment and binder percentage variations was studied with four samples, which screen printing pastes present the same concentration and color of conventional pigment (1% yellow) combined with TC blue 14, 11, 7 and 4%, samples K, L, M and N, respectively. Samples' spectral reflectance above 27 °C was measured in the spectrophotometer and a comparative analysis was conducted with color coordinates and CIELAB differences data, considering K as the pattern color. Furthermore, a similar analysis was conducted with samples which screen printing paste combines 1% conventional pigment yellow with equal concentration of TC pigments but different colors: 14% TC blue (K sample), 14% TC magenta (O) and a mixture of 5% TC blue and 9% TC magenta (P).

Light transmittance variation produced with the chromic behavior of K, O and P samples was analyzed through a photographic record and light transmittance measurement, with the setup described in the previous Sect. 3.2 for 1 m distance between sample and luxmeter.

Figure 3.5 presents the base framework of samples screen printed with 5% concentration of each pigment type, below and above 27 °C (left and right, respectively). The first raw presents the textile substrate on the left, followed by the samples screen printed with 5% TC pigment aligned by orange, red, magenta, blue and black colors. First column follows the same logic for samples with 5% conventional pigment, yellow, orange, red, magenta and blue, from top to bottom. Samples screen printed with mixtures of TC and conventional pigments were aligned in the matrix, according to the respective pigment versus color combination.

Through the framework analysis it was observed that by using the same concentration of both pigment types, samples screen printed with lighter conventional pigment colors attained a vast chromatic palette below 27 °C than samples with darker colors. For example, samples developed with conventional pigment blue attained shades in

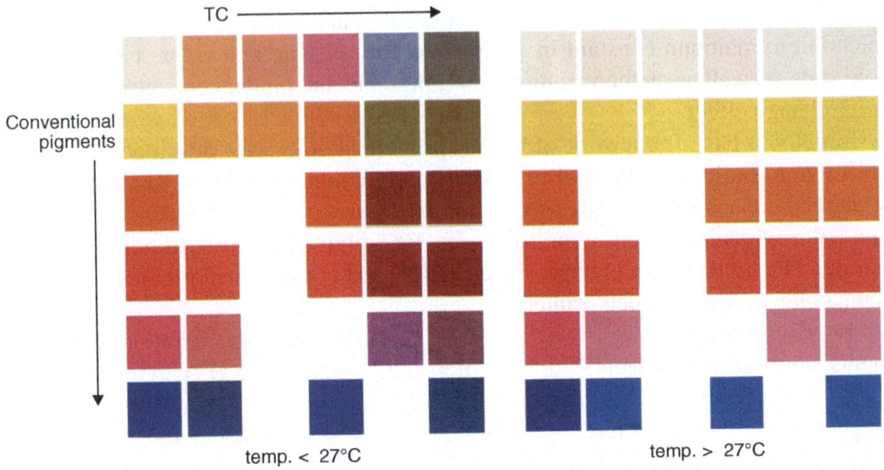

Fig. 3.5 Base framework of screen printed samples, below and above 27 °C (Cabral and Souto 2014)

the blue spectrum, whereas samples with conventional pigment yellow presented different colors. As the color strength of the conventional pigment demonstrated to be higher than the TC pigments, in order to achieve a vast combination of color change effects, the TC pigment concentration should be higher than the conventional, also considering the color(s) selected.

Above 27 °C, the TC pigments' residual color was observed in the first row samples through subtle color nuances, while the effect was barely perceived in samples with both pigments' types combined. However, as analyzed in Sect. 3.2, the incomplete decolorization of TC pigments also depends on their concentration, which is an important parameter to take into consideration, particularly when color change ratio from different to similar is to be designed, as will be further analyzed.

Figure 3.6 presents the framework of samples screen printed with a decreased ratio of conventional pigment yellow in relation to TC pigment increase, totalizing 10% pigments' concentration in each paste. Samples with TC orange, magenta, blue or black were organized in the framework from left to right columns. The framework picture on the left is below 27 °C and on the right is above.

With the decrease of conventional pigment concentration, from top to bottom in the framework arrangement, samples change from lighter to darker colors below 27 °C and the inverse occurs when heated. Although the changes rely mostly on color lightness, a transition between orange and red colors was observed in samples at the cold state, where TC magenta was combined (second column) and conventional pigment yellow varies from 2 to 1%. In samples with 0.5% conventional pigment and 9.5% TC pigment, the incomplete colorless state of the TC pigments was perceived.

The development of study samples with variation of pigments' concentration provided an important tool to understand color effects with the colourants handled and to aid an informed definition of color and concentration of the conventional

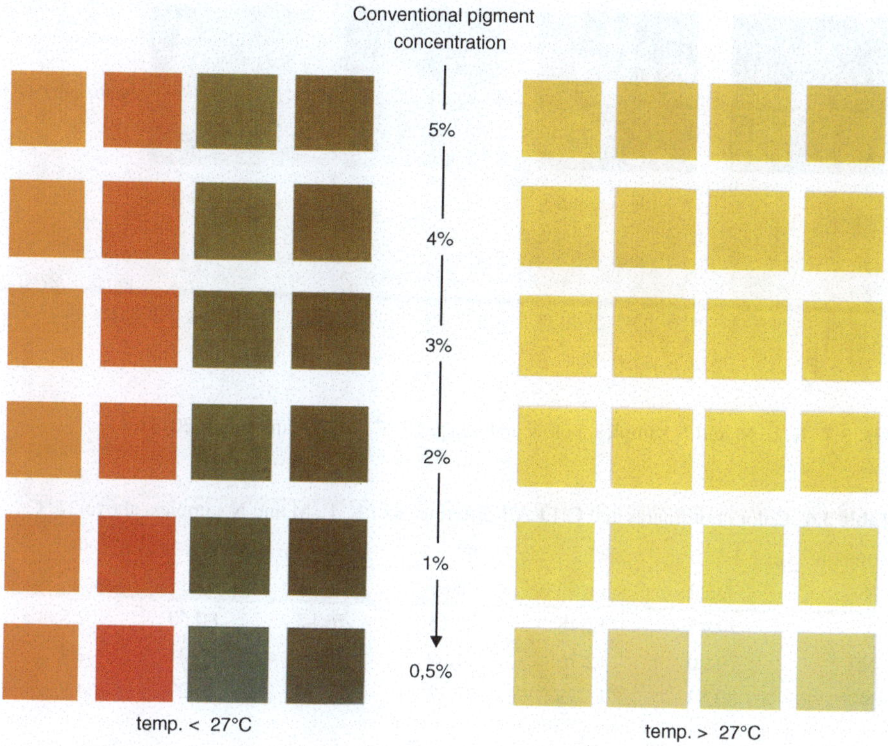

Fig. 3.6 Framework of samples with different TC and conventional pigments' concentrations, below and above 27 °C (Cabral and Souto 2014)

pigment to maintain constant in all samples, according to a color palette. For the following experiments, 1% of conventional pigment yellow was defined to apply in all pastes.

Within the color ratio focused in this study, TC and binder concentrations can also vary as well as more than one TC pigment color can be combined, to explore a vaster color set below samples activation temperature. However, these mixtures can also highlight color differences in samples' heated state, in terms of TC pigment residual color and binder concentration. To analyze how the variation of TC pigment and binder concentrations affect samples' colors, a study was conducted with K, L, M and N samples (screen printed with 1% conventional pigment yellow and TC blue at 14, 11, 7 and 4%, respectively). Samples' photos below and above 27 °C are presented in Fig. 3.7 and Table 3.6 presents the spectral reflectance data attained for measurements conducted with samples above 27 °C.

Below 27 °C, samples' colors vary from dark to light green with the concentration range from 14 to 4% of TC pigment blue. Above 27 °C, samples K, L and M appeared similar, while with N a lighter yellow color was perceived. Color measurement results with samples heated show that K is darker, less red and less yellow than the other

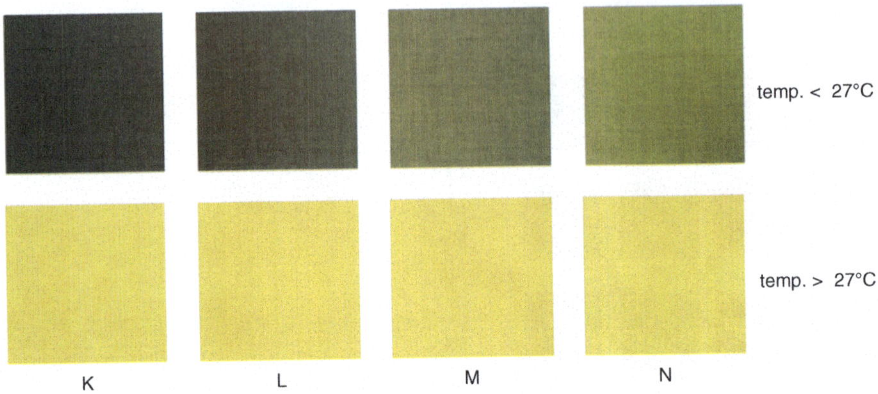

Fig. 3.7 K, L, M and N samples, below and above 27 °C (Cabral and Souto 2014)

Table 3.6 Color coordinates and CIELAB differences of K, L, M and N samples, above 27 °C

Sample	L*	a*	b*	C*	h	dE*
K	79.22	11.40	72.87	73.75	81.11	–
L	79.43	12.48	75.15	76.17	80.57	2.5
M	79.90	12.76	74.69	75.77	80.31	2.4
N	80.53	13.25	77.23	78.36	80.27	4.9

samples, with the highest values distance of 1.31 (L*), 1.85 (a*) and 4.36 (b*) in relation N. Considering sample K as the pattern color, CIELAB difference with N is approximately twice than in samples L and M. Moreover, while sample L is slightly lighter than M and thus more similar to K, its dE* is slightly higher. The opposite relation would be expected, with the higher difference of TC pigment concentration in relation to the pattern sample presenting a higher color difference due to the TC residual color effect. This result suggests that the incomplete colorless of the TC leuco dyes handled is not homogeneous along pigments concentration.

Furthermore, in regard to the samples developed and the color compared above 27 °C (yellow), the values attained between 14 and 7% TC samples were considered reasonable, considering that by direct observation the differences were not evident. The possibility to vary the TC pigment and binder concentration in pastes that change from different to similar colors extends the color palette range to be defined, below the TC pigment activation temperature.

To study how the combination of different TC pigment colors affects samples colors above 27 °C, a similar analysis was conducted with samples K (14% TC blue), O (14% TC magenta) and P (5% TC blue with 9% TC magenta). Samples' photos and spectral reflectance data are presented in Fig. 3.8 and Table 3.7, respectively.

By direct observation of samples' colors above 27 °C, K appeared darker and greener than the other samples with more relevant differences in relation to sample O. Color coordinates confirmed that K is darker, more green and less yellow than O

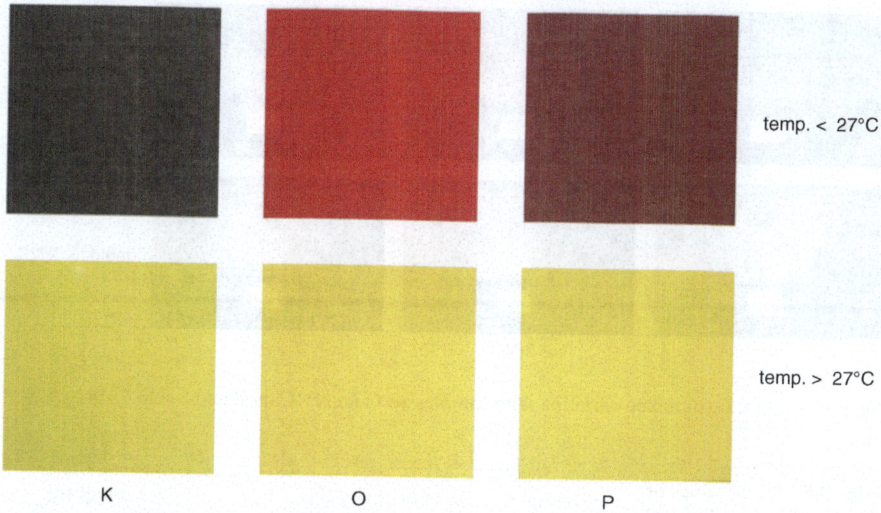

Fig. 3.8 K, O and P samples, below and above 27 °C

Table 3.7 Color coordinates and CIELAB differences of samples K, O and P, above 27 °C

Sample	L*	a*	b*	C*	h	dE*
K	79.22	11.40	72.87	73.75	81.11	–
O	80.06	16.80	76.07	77.91	77.55	6.3
P	79.79	15.22	74.37	75.91	78.44	4.1

and P samples, also showing that the high values distance take place in a* coordinate, attaining a difference of 5.44 with sample O and 3.82 with P. Sample P attained a lower CIELAB difference in relation to K than O, as expected because it presents 5% concentration of the same TC pigment color of the pattern sample, but it was still higher than samples developed with 11 and 7% TC pigment blue (L and M).

Samples K, O and P were also analyzed in regards to the light transmittance variation produced with the chromic effect. Figure 3.9, presents the photos attained at different temperatures. Below 27 °C, sample O presented the highest light intensity and the lowest was observed with K. Above 27 °C the light areas reflected on the wall appeared similar in size, however light tones were different, reflecting the influence of the TC residual color in the light transmitted, differentiated particularly by the warmer light tone with sample O.

Results of light intensity measurement (Table 3.8) demonstrate that below 27 °C, the highest light intensity difference is 0.46 lx representing 135.3% increase from K to O sample. Above 27 °C light intensity between samples is similar, being the highest difference of only 2% between K and P samples. Color change of sample K attained the higher luminosity difference.

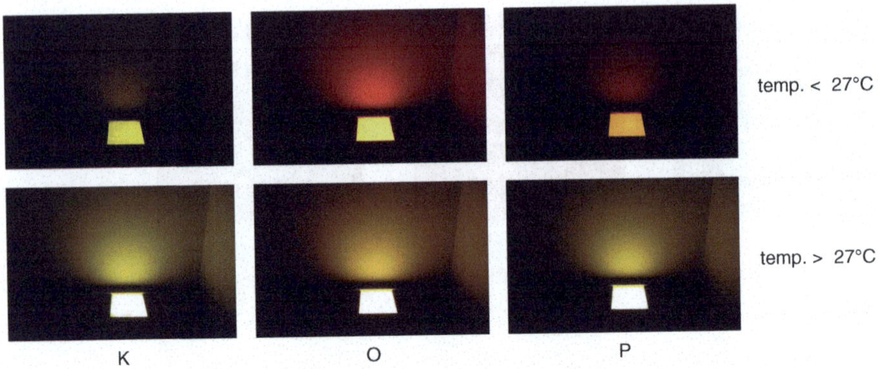

Fig. 3.9 Light transmittance variation with samples K, O and P (Cabral and Souto 2014)

Table 3.8 Mean values of light intensities with samples K, O and P

Samples	Light intensity (lx)		Luminosity difference (%)
	Temp. < 27 °C	Temp. > 27 °C	
K	0.34	3.01	89
O	0.80	3.04	74
P	0.42	3.07	86

The studies conducted in this section propose a process to develop TC textiles that change from similar to different colors. To achieve this color ratio, the color and concentration of the conventional pigment are independent variables in all screen printing pastes, whereas the TC pigment colors and concentration in relation to the binder are variable parameters. This study also allowed the creation of light transmittance variations from multiple light intensities to similar ones (Cabral and Souto 2014).

Moreover, it was found that the incomplete colorlessness of the TC pigments handled affected the colors' similarities with more relevance when using TC pigments of different colors than by applying the same pigment color (blue) at a concentration range from 4 to 14%. This effect, which was also perceived during light transmittance experiments, may be reduced with the addition of small percentages of conventional pigments of different colors, in order to homogenize the TC residual colors between samples, in the decolorized state. The possibility to match colors can be studied through paste recipe formulation processes, a method researched and presented in the follow section.

Furthermore, a sample with a geometric pattern was designed and screen printed with mixtures of TC and conventional pigments (K, O and P screen printing pastes), to demonstrate the defined color change ratio (Fig. 3.10). With the pattern sample developed, the differences between screen printed pastes colors above 27 °C appeared less obvious than comparing K, O and P samples, printed in solid colors.

Fig. 3.10 Sample: from different to similar colors ratio

3.4 Color Change: From Similar to Different Colors

In this section, the combination of TC and conventional pigments was researched towards the development of screen printed textiles that change from similar to different colors with temperature increase. The study conducted has researched optimization processes of screen printing pastes definition for TC textiles, by means of the colourimetric properties of the pigments handled, instrumental recipes' formulation and color assessment. These processes include conventional methods used by the textile industry in the colorimetry field (Gangakhedar 2010; Choudhury 2014a) and researched how they can be applied for the development of paste recipe formulation with TC pigments and mixtures of TC with conventional pigments.

In the first phase of this study, a set of screen printing samples was developed comprising of a defined concentration range for each pigment type and color. TC pigments were screen printed with 5–10%, concentration at 1% interval, totalizing six samples per each TC color handled: magenta, blue, black, red and orange. Conventional pigments were screen printed with 0.5; 1; 2; 3; 4 and 5% concentration, also summing six samples for each color: magenta, blue, black, red, orange and yellow.

After being screen printed and thermo set, the samples were measured in the spectrophotometer, at a temperature below 27 °C, following the procedures described in Fig. 3.1. The spectral reflectance results were used to create a database in Datacolor MATCH software, corresponding the samples' colourimetric properties according to colors and respective concentrations of each pigment type, defined as distinct colorant groups. This definition, allowed the formulation of paste recipes with different combinations of pigments' types and colors, given a pattern color and according to the color change effect aimed. The database created also included the textile substrate colourimetric data in samples' description.

To study the differences between colors of each pigment type, samples with 10% concentration of each TC pigment color were screen printed using the 50% PES and 50% CO substrate, measured and defined as pattern colors in order to formulate paste recipes with the database created, just selecting the conventional pigments type. Each sample was screen printed with the respective formulated recipe, thermo fixed and measured. The spectral reflectance results were compared with the correspondent pattern color through color coordinates and CIELAB differences data.

A study was conducted, which defined a mixture of two TC pigments as pattern color, to formulate paste recipes with mixtures of TC and conventional pigments that can change from similar to different colors. To systematize the specification of pigment versus color combinations, capital letters were used to define the pigment' type, TC for thermochromic leuco dyes and C for conventional pigments, while lower case letters represent the colors' first letter, has used in RGB or CMYK systems with capital letters (Choudhury 2014b). Black is described with letter k.

The samples developed in this study relied on the follow combinations: TCm + TCb; TCm + Cb and Cm + TCb. Sample Q combined 5.5% TCm with 5.5% TCb and after being developed and measured in the spectrophotometer, it was setup as the pattern color for instrumental recipe formulation, which defined the pigments' concentration for Q1 (TCm + Cb) and Q2 (Cm + TCb). According to the recipes obtained, each paste was elaborated and samples were screen printed with the same substrate as the concentration samples, dried and thermo fixed.

A comparative analysis of Q, Q1 and Q2 was conducted by direct observation of samples' colors and respective spectral reflectance measurement. Focused on color's similarities in the TC pigments' colorized state, the quantitative data presented respects to the colors measured with samples below 27 °C. Results were set in color coordinates and an overlap of spectral reflectance curves, which provides a visual profile of the color wavelength characteristics, showing the radiant power reflected at each wavelength of the visible light spectrum (Xin 2006; Setchell 2012). To analyze light transmittance variation through samples' color change, a photographic record and a light meter measurement were conducted with the setup described in Sect. 3.2, for 1 m distance between sample and luxmeter.

With the objective to study the possibility of improving similarities between samples' colors of formulated paste recipes in relation to the pattern color defined, an additional study was conducted. The process described for the development of samples Q, Q1 and Q2, was implemented for a new pattern sample R (TCm + TCb + TCk + Cy) and three formulated paste recipes: R1 (Cm + TCb + TCk + Cy), R2 (TCm + Cb + TCk + Cy) and R3 (Cm + Cb + Ck + Cy).

After samples' screen printing, thermo set and measurement, colors below 27 °C were analyzed by direct observation and by comparison of color coordinates in relation to the pattern color. The most distant values of L*, a* and b* coordinates were taken into consideration to perform adjustments in the previous recipes, by maintaining, adding or subtracting of small percentages of pigments. This study has included a paste recipe composed just with conventional pigments, to analyze the color matching effect in a combination that was expected to be more challenging, due to the use of only conventional pigments to reproduce a pattern color developed only with TC pigments.

Furthermore, this study has also encompassed a change of the textile substrate. The database previously created includes the data of the substrate used in the concentration samples. When defining the parameters for paste recipe formulation, substrate definition can be altered, allowing the formulation of recipes in regards to a new textile substrate color. For this study and production of a study prototype, the textile substrate 2 was applied (Fig. 3.1).

After the development of the optimized paste recipes, samples that attained greater similarity with the pattern sample were analyzed and selected for the development of a study prototype through rotary printing, which was conducted in an industrial context. Rotary printing process, involves the continuous rotation of pre-engraved cylindrical screens above the textile substrate and each printing paste is fed inside a cylinder, where a squeegee presses the paste out into the fabric (Miles 2003; Briggs-Goode 2013).

Previously to the textile rotary printing, pastes' viscosity was adjusted and printing parameters tested—rod squeegee size and magnetic field applied. With this study prototype, the effect of process change in color similarities was analyzed.

Figure 3.11 presents the photos of the concentration samples screen printed with conventional pigments (first row) and TC pigments (second row), decreasing concentration from the bottom upwards in each color set. Concentrations range selected for each pigment type, took in consideration color strength differences previously identified. However, with 0.5–5% concentrations for conventional pigments and 5–10% for TC pigments, relevant differences were still observed. Samples' spectral reflectance was measured and applied for the creation of a database with the colourimetric properties of the pigments handled.

The first experiment aimed to reproduce the colors of each TC pigment, in samples screen printed with conventional pigments, by formulating paste recipes through the developed database. By defining a TC sample as pattern color and selecting the

Fig. 3.11 Samples with different concentrations of conventional pigments (1st row) and TC pigments (2nd row)

software commands to reproduce it with the conventional pigment of the corresponding color, it was observed that the recipe options have attained high predicted dE* values for magenta and blue and no recipe was formulated for red and orange colors, due to the software limit of CIELAB difference 10 to formulate a recipe. Thus, each paste recipe was formulated with all conventional pigments' colors selected in the software commands, which resulted in recipe possibilities with lower predicted differences than previous recipes. For each TC pattern color, the three recipes with the lowest predicted dE* were selected to develop the samples. With TC orange as pattern color, there were just two options. Table 3.9 presents the samples' description with the respective paste recipes.

Screen printed samples are presented in Fig. 3.12. For each color group, TC pattern sample was placed on the left (vertical) and samples developed with conventional pigments on the right (horizontal). Table 3.10 presents the color coordinates and CIELAB differences attained between each study sample and respective pattern sample.

By a comparative analysis of the samples' colors and measurements results, it was possible to observe: (a) Sample magenta C presented great similarity with the TC pattern color and attained a dE* lower than the predicted, 1.0 instead of 1.9; (b) For blue, the lower color difference value was 3.9, with sample A. By direct

Table 3.9 Samples' description and respective paste recipes

Pattern colour (10% TC pigm.)	Study sample	Paste recipe with conventional pigments (%)						Predicted dE*
		Magenta	Blue	Black	Red	Orange	Yellow	
Magenta	Magenta A	1.617	0.009					0.3
	Magenta B	1.619	0.010					1.7
	Magenta C	1.561		0.003				1.9
Blue	Blue A		0.343	0.004	0.030			0.0
	Blue B	0.056	0.264	0.008				0.0
	Blue C		0.319	0.010				4.5
Black	Black A		0.130	0.262			0.043	0.0
	Black B		0.057	0.290				1.6
	Black C			0.309				2.1
Red	Red A				0.152	0.084		2.7
	Red B				0.224		0.075	4.3
	Red C	0.469			0.138			6.1
Orange	Orange A				0.116	0.452		5.8
	Orange B	0.279				0.539		8.1

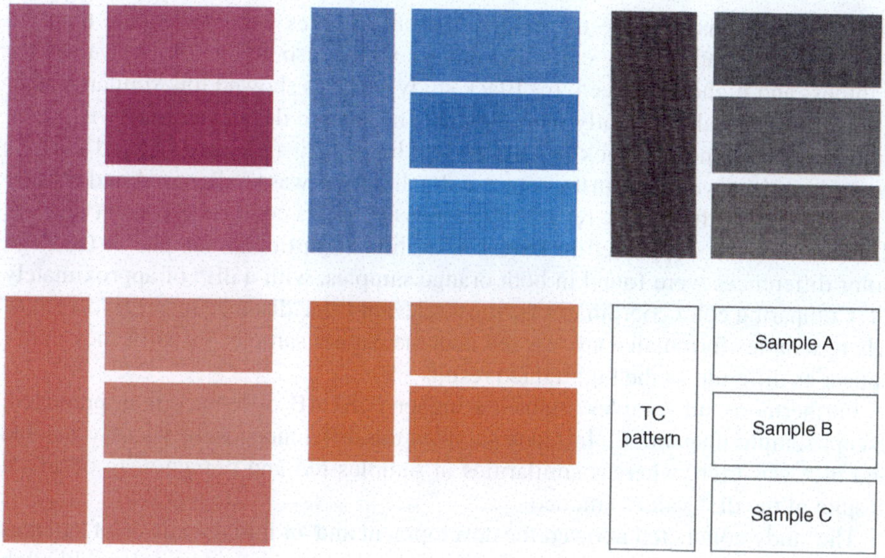

Fig. 3.12 TC pattern samples and conventional pigment samples

Table 3.10 Samples' color coordinates and CIELAB differences

Sample	L*	a*	b*	C*	h	dE*
10% TC magenta	53.60	45.29	−18.50	48.92	337.78	–
Magenta A	53.66	45.47	−19.47	49.46	336.82	2.4
Magenta B	48.62	49.16	−22.09	53.89	335.8	7.3
Magenta C	54.77	44.54	−17.54	47.87	338.51	1.0
10% TC blue	59.38	−5.84	−28.39	28.99	258.38	–
Blue A	56.88	−7.61	−32.14	33.03	256.68	3.9
Blue B	55.19	−7.15	−35.30	36.20	258.56	8.2
Blue C	57.73	−12.10	−40.07	41.85	253.19	13.3
10% TC black	36.66	−0.76	0.50	0.91	146.53	–
Black A	43.92	−2.31	−0.83	2.46	199.68	7.6
Black B	47.53	−0.21	−2.19	2.20	264.62	11.2
Black C	49.10	1.60	0.15	1.61	5.39	12.7
10% TC red	64.88	47.36	18.37	50.80	21.20	–
Red A	65.56	40.35	15.06	43.07	20.46	7.8
Red B	65.24	37.45	14.85	40.29	21.62	10.5
Red C	61.58	42.05	14.26	44.40	18.73	7.5
10% TC orange	63.20	56.60	42.08	70.53	36.63	–
Orange A	62.23	46.91	36.58	59.48	37.95	11.2
Orange B	59.91	47.47	37.50	60.49	38.31	10.7

observation it was possible to perceive that all samples were more blue than the pattern color, confirmed by color coordinates which also attained lower values for lightness and higher for green; (c) Black study samples showed low similarity with TC black, being all relevantly more lighter. The sample developed only with black conventional pigment—black C—attained the higher dE* (12.7); (d) All red samples were less red and yellow than the pattern color, lightness was similar for A and B while C was darker. Although the results of the measurements conducted present relevant differences, colors appeared to have similarities in particular sample C; (e) High color differences were found in both orange samples, with a dE* of approximately 11. Comparing color coordinates the most relevant value distance respects to a* axis, where samples formulated are less red than the pattern sample. Visually, they didn't appear as different as the highlighted results.

Furthermore, all samples attained a higher CIELAB difference than predicted, except sample magenta C. In black samples the differences were visually notable and dE* was high, whereas similarities in samples red and orange were observed despite of the dE* values attained.

The study conducted allowed the development and an initial analysis of samples whose color was defined by paste recipe formulations, given a pattern color. Although the results obtained have shown a low color similarity between the pigments handled, they have demonstrated the possibility to setup an initial paste recipe, which by experimental mixing would be more challenging to attain. It is also important to highlight that by direct observation and comparison of pastes composed by TC and conventional pigments (individually or combined), the color obtained after samples screen printing, drying and thermo set, is different. Thus, a direct matching of the pastes' colors is not reliable.

Moreover, while the preliminary analysis focused on paste recipe formulation only with conventional pigments, the objective of this study is to implement the database created for the development of textile color change behavior. Therefore, the pattern color does not have to be reproduced entirely by conventional pigments.

The second experimental study of paste recipes formulation through the database created, focused on the reproduction of a pattern color defined by a combination of two TC pigments, with samples that combine TC and conventional pigments. Pattern (Q) and samples with formulated pastes recipes (Q1 and Q2) are described in Table 3.11. Figure 3.13 presents the screen printed samples with the lower half heated above 27 °C.

Table 3.11 Paste recipes of samples Q, Q1 and Q2

Sample	Paste recipe (%)			
	TC magenta	TC blue	C magenta	C blue
Q	5.500	5.500		
Q1		6.128	0.616	
Q2	5.047			0.131

Fig. 3.13 Samples Q, Q1 and Q2, respectively

Table 3.12 Color coordinates of samples Q, Q1 and Q2, below 27 °C

Sample	L*	a*	b*	C*	h	dE*
Q	55.34	19.03	−27.74	33.64	304.45	–
Q1	54.92	15.54	−28.77	32.69	298.38	5.8
Q2	56.76	18.23	−30.28	35.34	301.06	4.2

By direct observation, samples appeared to have a resembling appearance below 27 °C, in particular comparing Q and Q1 samples, while Q2 appeared to be slightly more blue. When samples are heated, the TC pigment fades away and it is detected the lack of color on sample Q, the conventional pigment magenta on Q1 and blue on Q2, attaining the color ratio with similarities below 27 °C to different colors above it.

Comparing the samples' color coordinates (Table 3.12), pattern sample Q is lighter than Q1 and darker than Q2. Q is also more red and less blue than both samples, with a higher difference of values with Q1 for red and with Q2 for blue. Although visually the samples' color match appeared reasonable, CIELAB differences were higher than expected.

For a better understanding of color differences along the visible spectrum, samples' spectral reflectance curves were analyzed. The overlapping of spectral curves in Fig. 3.14, depicts the relation between the spectral reflectance at each wavelength of the three samples. At a temperature bellow 27 °C, Q2 had the blue higher reflectance (approximately 450–490 nm) and the lower reflectance on red wavelength (approximately 620–740 nm) while Q and Q1 curves are more alike. Although total color difference (dE*) of Q2 in relation to Q was lower than Q1, the spectral curves emphasized Q2 differences in specific wavelength ranges, being also visually more evident (Cabral and Souto 2012a).

Light intensity that passes through the textile samples was analyzed in regards to the color change ratio from similar to different, with temperature increase. Photos of samples' colors in light transmittance (Fig. 3.15) provide a visual confirmation of reasonable similarity between Q and Q1 below 27 °C, whereas Q2 is perceived as darker. Above 27 °C, samples' colors are distinct. Comparing the images of light

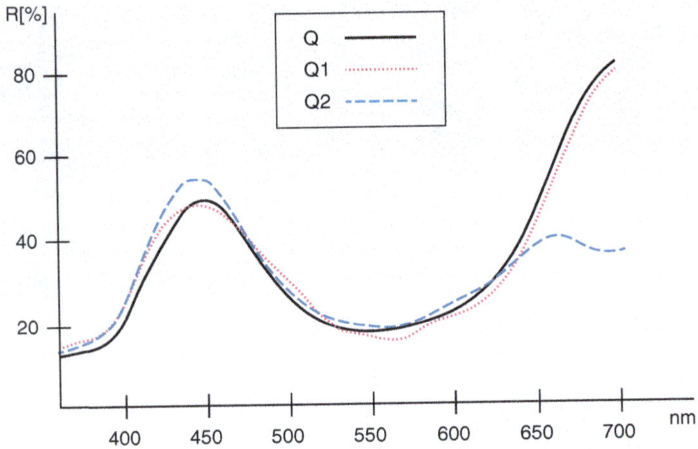

Fig. 3.14 Spectral reflectance curves of samples Q, Q1 and Q2, below 27 °C (Cabral and Souto 2012a)

transmittance variation with the light box perpendicular to the wall (Fig. 3.16), it was observed that below 27 °C, sample Q attained a slightly smaller area of light reflection on the wall than Q1 and Q2, while the tone of light was similar. When heated, the lighting areas between samples decrease from Q to Q1 and Q2, also presenting different colors.

Mean values of light intensity measurements (Table 3.13) also confirmed that whereas samples' colors appeared similar below 27 °C, light intensity varied. Sample Q attained the lower value with a difference of 078 lx from Q1, which is similar to Q2. When samples are heated, the higher difference observed was 3.32 lx between Q and Q2, representing 48% of light intensity difference. Upon temperature variation, the percentual luminosity difference increased from Q2, Q1 and Q (Cabral and Souto 2012a).

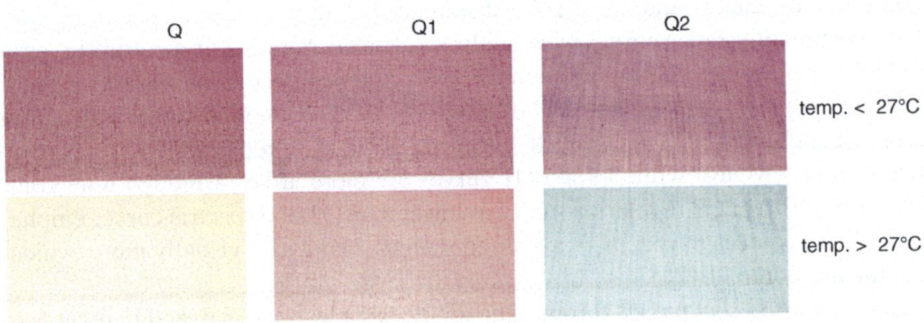

Fig. 3.15 Color of samples Q, Q1 and Q2 in light transmittance, below and above 27 °C (Cabral and Souto 2012a)

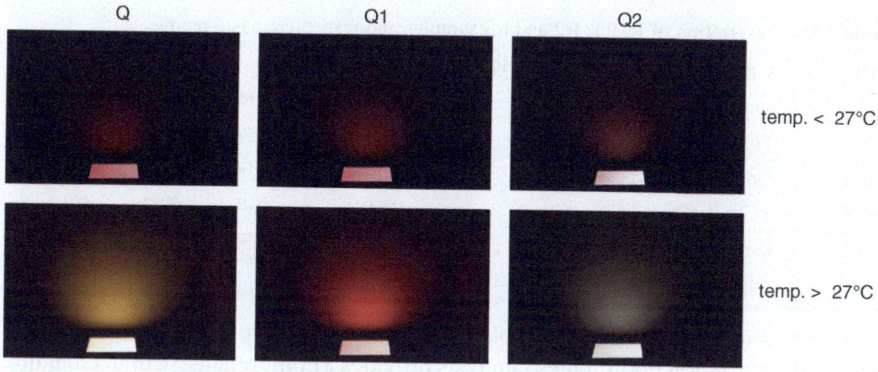

Fig. 3.16 Light transmittance variation with samples Q, Q1 and Q2

Table 3.13 Mean values of light intensities with samples Q, Q1 and Q2

Samples	Light intensity (lx)		Luminosity difference (%)
	Temp. < 27 °C	Temp. > 27 °C	
Q	1.23	6.86	82
Q1	2.01	4.24	53
Q2	1.96	3.54	45

The paste recipes defined for Q, Q1 and Q2 samples were applied for the development of a sample with a geometric pattern, in order to demonstrate the colors resemblance level below 27 °C by comparison of the color change in intercalated triangle areas, as well as to depict the color change ratio defined: from similar to different with temperature increase. Figure 3.17 presents the sample heated in the area on the right.

In despite of the reasonable comparison between the samples' colors below 27 °C, high resemblances were aimed. With the objective to study the possibility to improve color similarities, a third experiment was conducted where, after an analysis of the samples screen printed with the formulated paste recipes, color-matching adjustments

Fig. 3.17 Sample: from similar to different color ratio

Table 3.14 Description of R, R1, R2 and R3 samples and respective paste recipes

Sample	Pigments versus colors combination	Paste recipe (%)			
		Magenta	Blue	Black	Yellow
R	TCm + TCb + TCk + Cy	0.85	12.76	1.28	0.12
R1.0	Cm + TCb + TCk + Cy	0.14	13.00	0.05	0.12
R2.0	TCm + Cb + TCk + Cy	2.50	0.40	0.27	0.12
R3.0	Cm + Cb + Ck + Cy	0.20	0.35	0.01	0.13

were conducted based on the samples' spectral reflectance characteristics in the colorized state. Color coordinates' analysis provides a comprehensive understanding of the differences between formulated samples and pattern color. Along L*, a* and b* results, a relation between the pigments' concentration combined in paste can be approximately deduced.

This study was conducted with the textile substrate 2, which was measured in the spectrophotometer and the results were applied for the textile substrate definition in the software, during paste recipe formulations. Pattern sample R was defined with the follow setup TCm + TCb + TCk + Cy and samples with paste recipes formulated encompassed three combinations of pigments versus colors parameters. Table 3.14 presents the samples' description with respective paste recipes.

After samples development, a comparative analysis was conducted. Figure 3.18 presents the screen printed samples and the results of samples' color measurement were set in color coordinates and CIELAB differences, in Table 3.15. Through visual comparison, sample R1.0 showed chromatic resemblances with R, while R2.0 was perceived as more blue and R3.0 more green. Comparing color coordinates of samples with formulated paste recipes in relation to the pattern sample R, it was observed that: (a) R1.0 is lighter, more green and more blue, with the higher values distance between coordinates on b* (2.37 difference) and L* (1.56); (b) R2.0 is also lighter, more green and more blue, with the higher difference on b* (4.94) and a* (2.54); R3.0 is lighter, more green and less blue with high values distance on a* (7.36) and L* (5.48). The greater differences were found on the sample that combines four conventional pigments (R3.0), attaining 9.2 dE*. R1.0 was the sample which color was more alike to the pattern sample.

Fig. 3.18 Samples R, R1.0, R2.0 and R3.0, below 27 °C

Table 3.15 Samples R, R1, R2 and R3 color coordinates and CIELAB differences, below 27 °C

Sample	L*	a*	b*	C*	h	dE*
R	38.54	−7.27	−8.07	10.86	228.01	–
R1.0	40.10	−8.37	−10.44	13.38	231.26	3.0
R2.0	39.93	−9.81	−13.01	16.30	232.99	5.7
R3.0	44.02	−14.63	−7.15	16.28	206.03	9.2

Table 3.16 Samples' paste recipes adjustments

Sample	Pigments versus colors combination	Paste recipe (%)			
		Magenta	Blue	Black	Yellow
R1.0	Cm + TCb + TCk + Cy	0.14	13.00	0.05	0.12
R1.1		0.16	13.00	0.08	0.12
R2.0	TCm + Cb + TCk + Cy	2.50	0.40	0.27	0.12
R2.1		2.50	0.38	1.00	0.12
R3.0	Cm + Cb + Ck + Cy	0.20	0.35	0.01	0.13
R3.1		0.30	0.35	0.08	0.12

After analyze the differences between samples' color coordinates, addition, subtraction or maintenance of the pigments' concentrations was defined. The quantities to vary took in consideration the samples' value distance for each color coordinate in relation to R, the previous percentage applied for each pigment and the pigment type, as the conventional pigments present higher color strength than the TC pigments handled.

The first adjustments conducted in relation to the initial paste recipes are presented in Table 3.16, defined as follows: for sample R1.1 it was applied a concentration increase in Cm (0.02%) and TCk (0.03%); in R2.1, Cb was reduced (0.02%) and TCk increased (0.73%); in R3.1 there was a percentage increase for Cm and Ck (0.1 and 0.07%, respectively), while Cy was reduced (0.01%).

Figure 3.19 presents the new screen printed samples and respective pattern sample R. Table 3.17 presents the color measurement results of the new samples, in comparison to previous ones with the same pigments' combination setup. Samples screen printed with optimized paste recipes were more similar to the pattern sample than previous samples and attained a relevant decrease of CIELAB difference, new values range between 2.0 and 2.5. However, color differences were still perceived visually, particularly with R2.1 and R3.1 samples. Comparing the values for each color coordinate, all samples presented results more similar to the pattern sample except color opponent b* in sample R1.1 that was slightly more blue and R3.1 that was less blue than the previous sample.

Taking into consideration the improvement of colors similarity, additional cycles of paste recipes adjustments were performed. For each pigment combination, screen

Fig. 3.19 Samples R, R1.1, R2.1 and R3.1, below 27 °C

Table 3.17 Samples' color coordinates and CIELAB differences, below 27 °C

Sample	L*	a*	b*	C*	h	dE*
R	38.54	−7.27	−8.07	10.86	228.01	–
R1.0	40.10	−8.37	−10.44	13.38	231.26	3.0
R1.1	38.53	−7.00	−10.58	12.69	236.51	2.5
R2.0	39.93	−9.81	−13.01	16.30	232.99	5.7
R2.1	38.11	−8.75	−9.40	12.84	227.04	2.0
R3.0	44.02	−14.63	−7.15	16.28	206.03	9.2
R3.1	37.29	−8.41	−6.32	10.52	216.92	2.4

printed samples with the greater color similarity with R are presented in Fig. 3.20 and Table 3.18 presents the respective color measurement results.

Fig. 3.20 Samples R, R1, R2 and R3, below 27 °C

Table 3.18 R, R1, R2 and R3 color coordinates and CIELAB differences, below 27 °C

Sample	L*	a*	b*	C*	h	dE*
R	38.54	−7.27	−8.07	10.86	228.01	–
R1	38.35	−7.17	−9.17	11.64	231.96	1.1
R2	38.44	−8.77	−8.45	12.18	223.94	1.6
R3	38.23	−7.90	−8.77	11.80	227.98	1.0

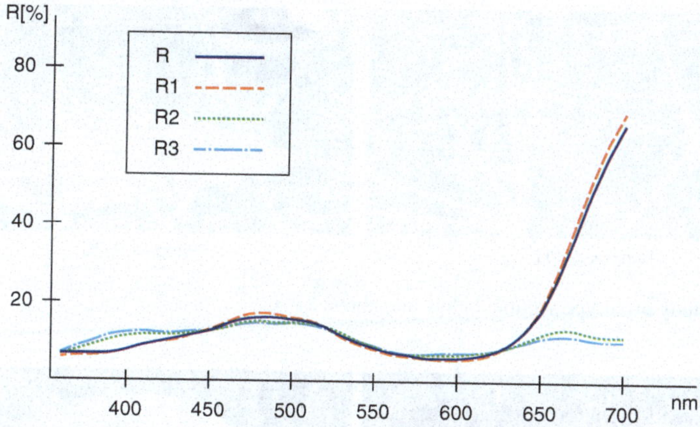

Fig. 3.21 Spectral reflectance curves of samples R, R1, R2 and R3, below 27 °C

By direct observation, samples' colors appeared similar to the pattern sample, although it was still possible to detect a slightly greener nuance in R2. Color coordinates confirmed the perceived resemblances and the highest value distance was 1.5 in a* coordinate of R2 sample. CIELAB differences with R1 and R3 were 1.1 and 1.0, respectively, and with R2 it was 1.6.

The overlapping of samples' spectral reflectance curves is presented in Fig. 3.21. R1 spectral curve was similar to R, while R2 and R3 presented relevant differences, both with higher reflectance in the violet band (below approximately 450 nm) and lower red reflectance (above approximately 620 nm). The high divergence within the red band, was also identified in sample Q2 in relation to Q and Q1 samples, highlighting differences between pigments' types in respect to the color blue: samples which pastes were elaborated with conventional pigment blue presented a lower reflectance within the red band, in comparison to samples' pastes developed with TC pigment blue.

Furthermore, while R2 and R3 curves were similar, by direct observation in natural light and comparison of color coordinates conducted for measurements with illuminant D65/10, R3 was more similar to R than R2. Regarding this divergence, by observing samples under different light sources, similarities in relation to R slightly vary with R2 and R3 samples. The illuminant metamerism (Wardman 2018) is particularly perceived with a subtle violet nuance in samples R and R1. In despite of this variation, samples' colors were still perceived to be similar.

A study prototype was developed in an industrial context through rotary printing process, with the same substrate and paste recipes defined for samples R, R1, R2 and R3. Pastes' viscosity varied between 23, 27, 45 and 50 dPa from R to R3, respectively and a thickness agent was applied to present 90 dPa in each paste. Rotary printed study samples were developed with different rod-squeegee numbers (differing in diameter and weight) and magnetic field applied, to analyze colors and select the parameters to apply in the study prototype printing. The study prototype accomplished 22.0 ×

Fig. 3.22 Study prototype section, below and above 27 °C

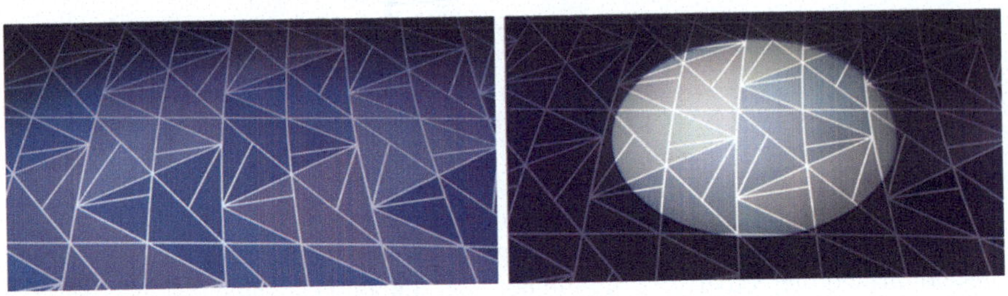

Fig. 3.23 Study prototype section, under two different artificial light sources

1.5 m, developed with a geometric pattern composed by four cylinders with 91.4 cm rapport.

Figure 3.22 presents a photo of a section of the study prototype, below 27 °C (image on the left) and with an area heated above 27 °C (image on the right). Below 27 °C, the study prototype colors have shown a decrease in similarity in relation to R samples' colors. Also being more obviously perceived is the colors metamerism as illustrated in Fig. 3.23, which presents a part of the study prototype illuminated with two different artificial light sources.

Reflecting the effect of the pigments' application process on colors, the results attained highlight the importance of creating the software database with concentration samples developed with the same process and respective parameters to apply in final products.

3.5 Conclusions

Throughout the research discussed in this section, textile chromic effect on light transmittance was analyzed, TC pigments behavior and properties were studied and two process proposals were researched. Paste recipes were elaborated with TC and conventional pigments to screen print textiles that change color according to predefined ratios.

TC textile behavior as dynamic light filter entails a set of dependent variables for the designer to explore. The preliminary analysis allowed an initial understanding on the relation of light transmittance and TC pigments' color, concentration and application processes. According to pigments absorption of specific ranges of light wavelengths to reflect or transmit their colors, samples screen printed with different TC pigment colors at the same concentration attained different luminosity differences through the chromic behavior. By changing the TC pigments concentration in the paste, the results demonstrated the possibility to create ratios of light transmittance variation. For example, 10% TC red (sample B) attained similar percentual luminosity difference than 1% TC black (sample E). Besides the quantitative data, perception of the light change in a dark room was evident, regarding intensity levels but also with great emphasis on light tone transition.

Application processes of the colorant in the substrate also affected the light transmitted and samples' appearance. By using the same pigments' colors and concentrations, the sample developed with TC and conventional pigments mixed in paste (H) attained higher luminosity difference than the sample where individual pastes for each pigment were screen printed through a layer overlapping (I). On the other hand, the two pigments' layers sample with TC on top (I) attained a darker expression, which for applications not focused on light transmittance purposes, present the possibility to use less TC pigment or explore darker color change effects with similar pigments' quantities than by mixtures in the same paste. The combination of dyeing and screen printing processes have also shown great potential to explore color change effects.

Regarding the combination of TC and conventional pigments in the screen printed paste, the process outlined for the development of TC textiles that change from different to similar colors with temperature increase, demonstrated that color and concentration of the conventional pigment (or combination of conventional pigments) are fixed parameters, whereas the TC pigment colors and concentrations are variable parameters. Samples with the chromic ratio defined varied light transmittance from heterogeneous to similar luminosities.

To screen print TC textiles that change from similar to different colors, the process developed applied instrumental methods of paste recipe formulation. The database created with the colourimetric properties of the conventional and TC pigments colors, below their activation temperature, provided the possibility to formulate paste recipes where colors of each pigment type are selected as distinct colorant groups. This individual selection enabled the definition of pigments' colors in respect to their dynamic or static behavior, depending on the color change effect aimed. Samples with similar colors in relation to a defined pattern sample were developed, through one paste recipe attempt for each pigments' combination. Color resemblance below the TC pigments activation changed to different colors with temperature increase, extending similar light transmittance to heterogeneous luminosities.

However, the color match of a pattern sample with samples which pastes have the opposite pigment type was found to be challenging to attain, an issue also affected by the pigment colors' number in the paste and the combination with specific colors.

Thus, an additional process of paste recipe adjustment was proposed to be conducted after the development of paste recipe formulation, based on a comparison of samples' color coordinates. Samples developed attained CIELAB differences of approximately 1.

The optimization of paste recipes formulation and screen printing processes developed, provide a tool to explore color change ratios across a wide range of color palettes and textile applications. These processes apply some base procedures that are well known by the industry, which can enhance a path between TC textile design and manufacturing, through the systematization of procedures working with TC pigments and the color match within textile collections. Nevertheless, the designer must fully understand the TC chromic behavior in order to define colors and select the appropriate commands in the software to attain the desired color change effect.

In the experimental work conducted, it was observed that the TC pigments handled presented considerably low color strength in comparison to the conventional pigments as well as low color similarity between pigment types. Moreover, the TC pigments' transition from colored to colorless with temperature increase was incomplete. A residual color was perceived in the pigments' decolorized state, depending on colors and concentrations applied. With samples screen printed with different colors of 14% TC pigment, more obvious differences between residual colors were perceived than in samples with the same pigment color at different concentrations for a 10% interval (4–14%).

Furthermore, colorfastness of the pigments handled was assessed, having the TC pigments presented similar or greater stability to rub and wash in relation to the conventional pigments, but light fastness was low: after 8 h of UV light exposure, TC samples obtained CIELAB differences from 7.2 to 12.6, in relation to samples' colors before light exposure. Regarding this research objective to work with artificial light in interior spaces, poor light stability would not directly affect TC behavior lifespan. However, this is a major issue for a broad range of applications and, although research have been conducted on this topic, there are still great limitations (Ibrahim 2012; Ferrara and Bengisu 2014).

The research conducted in this section and the results attained through the samples developed and respective behavior analysis, provides an understanding to design TC textiles and set up base procedures to implement in future developments in Dynamic Light Filters research (Chap.7).

References

Briggs-Goode A (2013) Printed textile design. Laurence King Publishing, London

Cabral I, Souto AP (2012a) Colour spread: light, thermochromics and conventional pigments. In: Dragcevic Z (ed) 6th International textile, clothing & design conference, 2012, Dubrovnik, Croatia

Cabral I, Souto AP (2012b) Thermochromic filters effect on static light. Int J Vis Des 6(2):55–64. https://doi.org/10.18848/2325-1581/CGP/v06i02/38714

Cabral I, Souto AP (2014) Thermochromic textile structures: a dynamic ambient light design. Int J Des Objects 7(3):23–35. https://doi.org/10.18848/2325-1379/CGP/v07i03/38683

Choudhury AKR (2014a) Principles of colour and appearance measurement: visual measurement of colour, colour comparison and management, vol 2. Woodhead Publishing, Cambridge
Choudhury AKR (2014b) Principles of colour appearance and measurement Volume 1: Object appearance, colour perception and instrumental measurement. Woodhead Publishing, Cambridge
Descottes H, Ramos CE (2011) Architectural lighting: designing with light and space. Princeton Architectural Press, New York
Ferrara M, Bengisu M (2014) Materials that change color: smart materials, intelligent design. Springer International Publishing, Cham
Gangakhedar N (2010) Colour measurement methods for textiles. In: Gulrajani ML (ed) Colour measurement: principles, advances and industrial applications. Woodhead Publishing, Cambridge
Ibrahim W (2012) An investigation into textile applications of thermochromic pigments. PhD thesis, Heriot-Watt University
Miles LWC (2003) Textile printing. Merrow Publishing Co. Ltd., Watford
Setchell JS (2012) Colour description and communication. In: Best J (ed) Colour design: theories and applications. Woodhead Publishing and Textile Institute, Oxford
Viková M (2019) Type of chromic materials. In: Viková M (ed) Chromic materials: fundamentals, measurements, and applications. Apple Academic Press, Waretown
Wardman RH (2018) An introduction to textile coloration: principles and practice. Wiley, Hoboken
Worbin L (2010) Designing dynamic textile patterns. Doctoral thesis, University of Borås
Xin JH (2006) Total colour management in textiles. Woodhead Publishing and The Textile Institute, Cambridge

Chapter 4
Electrical Activation of Thermo-responsive Textiles

Defined by the ability to conduct electrical current, conductive materials are a key topic in smart textiles field, as they enable the transfer of energy or data and impart textiles with electronic functionalities. Electrical conductivity can be conferred to textiles at the fiber or filament level, threads, substrates and applied finishings. Advances in this area have been providing a wide range of ready-made metal-based materials with increasingly textile-like mechanical and tactile properties.

Designing thermo-responsive textiles, conductive materials are enabled to produce temperature variation through electrical activation, according to their properties, integration in the substrate and thermal conductivity of the textile structure. In this research, the integration of conductive materials in textiles was studied with the objective to activate color change of TC textiles.

This chapter starts with an overview on electroconductive materials, fabrication technologies associated with the development of electroconductive textiles and applications, with particular emphasis on textile heating systems. The experimental work developed with metal-based conductive threads and pigments is discussed, including their integration in woven substrates and the analysis of their electrical and expressive qualities for resistive heating activation of color changing textiles.

4.1 Electroconductive Materials

Defined by the ability to conduct electrical current, electroconductive materials are a key topic in smart textiles field, as they enable to transfer energy or data and impart textiles with electronic functionalities (Eichhoff et al. 2013; Chatterjee and Maity 2018). Cutting-edge potentials of fiber-based materials and products embedded with technology and added functionalities are also intertwined with great challenges. Functional feasibility of materials and devices that were traditionally rigid and heavy, combined with processability and comfort demands of soft and flexible

I. D. Cabral et al., *Dynamic Light Filters*, Springer Series in Design and Innovation 3, https://doi.org/10.1007/978-3-030-39529-2_4

textile structures have been great concerns, in parallel to wear and washing stability (Dias and Ratnayake 2015; Kumar and Vigneswaran 2016).

Enhanced means for conductive textiles development have undergone crucial advances, namely the emergence of conductive materials with increasingly textile-like mechanical and tactile properties (Kirstein 2013; Koncar 2016). As advances into new enabling technologies are moving towards more seamless and unobtrusive integration of electronics into textiles, creative approaches are also exploring new design perspectives considering functional and expressive performances (Pailes-Friedman 2016; Schneegass and Amft 2017).

Conductive textiles hold a wide range of applications and electrical conductivity requirements, for example surfaces with low conductivity are applied for antistatic purposes, whereas for signal transmission, high conductivity is desired (Schwarz and Langenhove 2013; Zeng et al. 2014). Encompassing diverse electrical, mechanical and visual properties, conductive material selection is an important consideration for conductive textiles development including their manufacturing, functional and expressive capabilities (Tao and Koncar 2016; Varga 2017).

Electrical current that passes through a conductive material is defined by the flow of electrical charges, which are commonly carried through electrons. To generate this motion, an electrical voltage is applied between two points of the conductor, creating a potential difference that forces the charges to flow. In addition, conductive materials present different resistance levels to the flow of electrical current depending on their properties and dimensions (Storey 2009; Schwarz and Langenhove 2013).

Ohm's law expresses this relation mathematically, where for a defined voltage (V) applied, the electrical current (I) that flows though the conductor is directly proportional to its electrical resistance (R): $V = I \times R$.

Accordingly, low resistance materials are highly conductive, whereas high resistance results in lower conductivity. The unit of electrical current is Ampere (A), being 1 A approximately 6242×10^{18} electrons (1 C) per second passing through a point in the conductor. Voltage is measured in Volt (V) and electrical resistance in Ohm (Ω).

Electrical conductivity is conferred to textiles by means of metals, intrinsically conductive polymers (ICP) and carbon-based materials. Metals or alloys present high levels of electrical conductivity, although they have low elasticity and are heavier than conventional textile fibers. Common metals used for textile integration are copper (Cu), silver (Ag), aluminum (Al), stainless steel and nickel-chrome (NiCr) (Schwarz and Langenhove 2013; Cork 2015; Kumar and Vigneswaran 2016). ICPs such as polyaniline, polyacetylene, polypyrrole and poly(3,4-ethylenedioxythiophene):polystyrenesulfonate (PEDOT:PSS) are lightweight, flexible and can also be blended with other materials during processing. Ongoing research in this field has been particularly focused on these materials' electrical properties and electronic possibilities, as well as processability, environmental stability and toxicological challenges (Zeng et al. 2014; Qu and Skorobogatiy 2015). Carbon nanotubes and graphenes are carbon-based conductive materials that have demonstrated great potential for textile applications. At the research stage, their major advantages rely on being lightweight and presenting high mechanical strength (Cork 2015; Miao 2015).

4.2 Electroconductive Textiles

Manufacturing processes of electroconductive textiles can be conducted at different levels, from fiber or filament to threads, substrates and finishings. Involving conventional and innovative textile processes, fabrication technologies of conductive textiles are based on two major approaches: textiles entirely produced by intrinsically conductive materials and textiles blended or specially treated to gain conductivity (Schwarz and Langenhove 2013; Cork 2015).

Conductive fibers include synthetic materials filled with conductive particles, textile fibers coated with conductive materials and fibers or filaments produced entirely by conductive materials, such as metal monofilaments produced by wire drawing. These fibers and filaments are applied for conductive threads manufacturing, being combined with conventional textile materials, or used directly in substrates processing, for example in the production of conductive nonwovens (Xue et al. 2005; Stoppa and Chiolerio 2014; Guo et al. 2016).

Metal monofilaments wrapped around a textile-based core thread assume a spring shape, where the twist number defined can vary to tailor the threads' electromechanical properties. Another approach is to spin metal wires with non-conductive filaments, setting one or more conductive strands in the plied thread. In both cases the textile material acts as the load carrier, having higher mechanical strength and flexibility than the metal filaments and accounting for the thread processability and handling. Textile threads can also be plated with a conductive layer. In this case, the overall surface is conductive but it is also more susceptible to cracking. To reduce conductivity issues, this method is commonly applied for the development of multifilament threads (Eichhoff et al. 2013; Locher 2013; Rakshit and Hira 2014).

Conductive fiber, filaments or threads are integrated into the textile structure or applied onto the substrates' surface. Through integration processes such as weaving, knitting or braiding, conductive materials are at the same time functional and structural elements of the textile. Their patterning possibilities depend on the textile construction where diverse relations between form and function can be explored. In woven substrates, conductive threads are mostly used in the weft, due to greater flexibility for the circuit design and threads exchange. Conductive threads in the warp are also avoided due to the higher stress that they are exposed during processing and consequent increased possibility to affect or lose conductivity (Eichhoff et al. 2013; Stoppa and Chiolerio 2014). Veja (2014) researched weaving methods for the integration of conductive materials and electronic devices, highlighting in particular the potential of using double cloth techniques; inlay weft weaving; and manipulation of floats for the development textiles with multiple functions.

Embroidery or sewing processes allow a higher flexibility for the conductive threads pattern definition, as the integration on the surface does not depend on the textile structure. E-broidery (electronic embroidery) technologies have been explored in many applications, taking advantage of precise design control and stitched circuit pattern through CAD (Computer-Aided Design) software (Stoppa and Chiolerio 2014; Li and Tao 2015). The Forster Rohner company patented an e-broidery technology that allows the integration of conductive threads and active components during production (Bosowski et al. 2015; Pailes-Friedman 2016).

Finishing processes also promote the application of conductive materials on the textile surface. Conductive coating techniques include physical and chemical vapor deposition, electroplating, electroless plating, among others (Montazer and Harifi 2018). Electrical conductive properties of coated fibers, threads and substrates can be tailored through the coating layer thickness and the conductive materials selected (Shim 2010; Wang et al. 2010; Zeng et al. 2014). Printing is a versatile method to confer conductivity in textiles, enabling the creation of a desired pattern in localized areas, using conductive materials in paste form. Screen and inkjet printing are the main techniques used, presenting the advantage of low material waste and the possibility of either large-scale or low volume fabrication (Kirstein 2013; Yang et al. 2013; Torah et al. 2015).

Critical challenges of coated and printed conductive textiles rely on: the textile surface roughness, which interferes in the conductive layer uniformity and thus electrical performance; curing temperature, considering the textile substrates' threshold limits; physical and chemical properties, in particular the change or loss of electrical conductivity as the result of abrasion, bending and washing conditions (Stoppa and Chiolerio 2014; Li and Tao 2015; Torah et al. 2015). Research focused on this topic encompasses development of conductive pastes with enhanced properties such as stretchability (Matsuhisa et al. 2015; Jin et al. 2017) and processes to seal the conductive printed layer, involving the production of additional protective layers, commonly applied with polyurethane-based pastes by coating, lamination or screen printing (Yang et al. 2013; Yokus et al. 2016). The sealing consists of an interface layer applied in the substrate for waterproofing and to create a smooth surface for the conductive paste and an encapsulation layer for electrical insulation and external protection. Although still not optimal, improved results in flexibility and durability to abrasion, washing and dry cleaning have been attained (Kazani 2012; Gordon et al. 2014; Li and Tao 2015; Jin et al. 2017).

Initial uses of metal-based conductive materials in textiles did not encompass functional applications. Indian metal silk organzas are woven substrates that include silk threads wrapped with thin metal films and provide textiles a brilliant appearance. Regarding aesthetic and social status purposes, embroidery with noble metal-based threads have also been applied in monarchic and ecclesiastical apparel (Schwarz and Langenhove 2013; Cork 2015). Contemporary approaches of metal-based textiles have been eminent in the work of Reiko Sudo (Nuno Corporation), namely by exploring technologies not traditionally associated with textiles, in which plating techniques of the automotive industry are an example, inspiring new visions for textile design (Nuno n.d.).

Within functional aspects, conductive textiles present a wide range of applications. Electrically conductive pathways enable the transfer of energy or data between interconnected electronic devices or other conductive materials, such as thermo-responsive SMAs (Rakshit and Hira 2014; Qu and Skorobogatiy 2015). In parallel, conductive materials' properties and structures can be devised to enable electronic functionalities. Textile-based systems may comprise of sensing and actuation behavior, wireless communication, data processing and energy storage or harvesting (Eichhoff et al. 2013; Langenhove 2015; Toprakci and Ghosh 2015; Wang et al. 2015; Wilson and Mather 2015). Permeated through nanotechnology advances and electronics miniaturization, emerging research aims at the development of textile-based electronics, also including the potential of micro-electronics integration into fiber forms (Dias and Ratnayake 2015; Varga 2017).

To create a textile electronic system, the conductive circuit has to be connected to a power source. The technology and interface applied for the elements' link must assure a reliable electrical connection with low electrical resistance (Suganuma 2014; Mehmann et al. 2017). Joining technologies for conductive textiles encompass permanent or reversible connections. Methods used for fixed connections include soldering (with lead-free solders), crimping, bonding (with conductive adhesives) and also conventional textile processes such as embroidery or sewing. Reversible joining can be produced with the inclusion of conventional or textile-based elements. Some examples include pin headers, metal eyelets, conductive zippers, conductive hook and loop fasteners (Velcro) (Locher 2013; Mecnika et al. 2015).

According to the conductive textile characteristics and applications, electrical insulation of the connectors and conductive elements may also be required. As previously described, encapsulation layers developed through printing or coating processes with insulator agents can provide this property (Li and Tao 2015; Jin et al. 2017). Other techniques include embroidery with conventional threads, iron-on patches, adhesive films and ribbons. One major challenge in the development of connections and respective insulation is to achieve stable performance and textile flexibility (Lin 2015; Kumar and Vigneswaran 2016).

The application of conductive textiles as sensors and/or actuators involves the transmission and also transformation of energy or data. Textile sensors are able to transform physical stimuli, such as pressure, temperature or humidity, into processable electrical signals. This transformation lies in the change of conductive materials' electrical properties with a specific input. As actuators, conductive textiles convert electrical signal inputs into physical responses, for example light or heat (Langenhove 2015; Toprakci and Ghosh 2015). Thermal variation of conductive materials is based on the process by which electrical current passes through a conductive material and, according to their electrical resistance and voltage applied, power dissipates as heat produces a temperature rise—the Joule effect, also known as Joule or resistive heating. Conductive materials as thermal actuators are also called resistive heaters (Storey 2009; Mbise et al. 2015; Torah et al. 2015).

Textile heating systems are usually developed to provide safe and comfortable body thermoregulation. An initial example is E-CT from Gorix, a patented fabric that was used in the production of heated boots, gloves, among others (Baurley 2005).

Textile heating in car seats is an established application, where the electrical current that passes through the conductive materials is controlled within the possibility to select different heating levels. Nichrome monofilaments and FabRoc® carbon loaded silicone threads have been reported for this application (Mbise et al. 2015). Performance outwear, pant liners, vests and scarfs are commercially available examples of heated garments and accessories (Wagner 2013; Tong and Li 2015; Kumar and Vigneswaran 2016). Whereas some applications use heating pads to enable the garments' washing without the conductive textiles, there are already marketed solutions that withstand machine-washing. WarmX® is a brand of self-warming underwear and base-layer garments that use silver-plated polyamide threads in knitted structures. They use a mini power-controller to heat up the textile, being the only element disconnected for washing (WarmX n.d.; Pailes-Friedman 2016).

Besides body thermoregulation, resistive heating is also commonly applied for temperature control of TC textiles. As discussed in Sect. 2.4 , the heating elements that activate TC color change have a significant effect on the textile dynamic behavior and expressions. Thermal variation induced by conductive materials integrated in the textile substrate involves an intrinsic relation between the TC textile and the conductive materials' properties, patterning and fabrication processes. Thus, the physical production of the conductive textile and respective electrical properties are crucial considerations in smart textile design.

The use of conductive materials to activate textile color change is based on temperature increase, while the textile cooling is not commonly electrically controlled. Hu et al. (2005) researched the development of a cooling fabric with conductive polymer coated fibers, based on the 'Peltier effect'—the effect of heat production or absorption at the junction of two dissimilar metals when electrical current passes through their junction (Gibilisco 2001; Wang et al. 2010). Although a cooling effect was achieved, the performance was limited also attaining electrical degradation of the polymers. In collaboration with European Space Agency, Grado Zero Espace company developed a textile cooling system adapted from astronaut suits, involving a complex structure of thin tubes for the circulation of a liquid that was cooled by a Peltier element. The cooling system was installed in a F1 pilot racer suit, providing thermo-regulation (Espace n.d.; Langenhove 2015).

Future possibilities to integrate cooling effects in thermo-responsive textiles may provide a meaningful tool to design chromic and shape memory textiles behavior. At the research stage, the challenges involved in the integration of cooling systems in textiles, namely structures' complexity, weight and non-flexibility, still limit potential applications (Langenhove 2015; Torah et al. 2015). Considering this research aims, the integration of tubing structures and Peltier devices also would compromise textiles' light transmittance. Thus, the use of controlled cooling systems was not considered.

In this research, the integration of conductive materials in textiles substrates aims to provide a means for electrical activation of color and shape change through resistive heating. For color change, the conductive pattern will be the thermal actuator. Nitinol alloys are conductive materials and shape change will be produced through their resistive heating properties. The use of conductive materials for shape change activation will establish the electrical connections between Nitinol segments and electronic components. Furthermore, within a great range of materials and properties, conductive materials' electrical properties differ, enabling to explore not just functional performance but also textiles' expressive qualities.

4.3 Experimental Work with Metal-Based Conductive Materials

Given the research aim to activate color and shape change textiles through resistive heating, this section studies the integration of conductive materials in textile substrates and presents an initial analysis on their resistive heating behavior.

For the development of this research phase, a conductive materials catalogue was built comprising of materials that conduct electrical current and can be integrated in textile substrates. Swatches of the materials collected were organized by type: fibers, threads, yarns, ribbons, tapes, fabrics and printed pigments. In each section type, swatches of the same brand were grouped and brief description was included.

The initial purpose of this catalogue was to assemble a list of the conductive materials gathered, providing a visual and tactile analysis, also allowing the inclusion of additional samples. The concept of collection was addressed. Collection as a set of items of a specific class and collection as an act of gathering together, a work in process that can continuously grow and expand (Collection n.d.). Within this perspective, the catalogue scope has evolved to a collection of data attained in the experiments with the conductive materials, focused on resistive heating activation of thermo-responsive textiles.

In each experiment, the body of results involved quantitative and qualitative data with a range of interdependencies. Conventional methods of data assembling by type or category were not found to be comprehensive for an analysis of the materials' properties and behavior, thus data documentation alternatives were explored.

Regarding abstract data, Stamatellos (2007) defined the concept of collection in computer science as "a grouping of some variable number of data items that have shared significance to the problem being solved and need to be operated upon together in some controlled fashion.". The main idea of data relations and influence in a specific phenomena or problem, lead to the logic behind the data organization conducted in this research documentation. The focus was not data types, rather the relations between dependent variables, according to the objective of each experiment.

The first section of the experimental work concerned with conductive threads integration in plain weave structures and studied their electrical resistance properties for textile color change activation. In the second experimental section, a similar study was conducted with conductive pigments, focusing on the screen printing parameters influence in the printed traces electrical resistance and consequently resistive heating effect. In addition, electrical resistance of each integrated material type was also analyzed with the textile samples folded.

Materials, equipment and methods applied in this chapter experimental work are described in Fig. 4.1. Samples description is presented in each sub-chapter.

MATERIALS

Conductive threads:

CT1 - Bekinox VN 12/1x275/100Z	CT15 - Shieldex 110/34 2ply HC+B
CT2 - Bekinox VN 12/4x275/100S	CT16 - Elinox SPP 35 300Z
CT3 - Bekintex BK 50/2	CT17 - Elinox PES HT1100 + VN60 400 tS/Z
CT4 - Bekintex BK 50/1	CT18 - LessEmf silver plated nylon
CT5 - Bekintex VN 14/2X90/175S/HT/PFA	CT19 - Lamé LifeSaver
CT6 - Bekiflex CA 63/8X7/80S/0.245/PFA	CT20 - Linox
CT7 - High Flex 3981 7X1 Silver	CT21 - Sparkfun stainless steel 2ply
CT8 - High Flex 3981 7x1 Kupfer Blank	CT22 - Sparkfun stainless steel 4ply
CT9 - Constantan High-Flex 8394 7×1	CT23 - Copper monofilament 100 μm
CT10 - Elitex 235/f34 PA/Ag	CT24 - Insulated copper monofilament 160 μm
CT11 - Elitex GarnArtTPU667_235/f34 PA/Ag	CT25 - Stainless steel monofilament 50 μm
CT12 - Elitex 110/f34/2ply PA/Ag	CT26 - Nichrome monofilament 200 μm
CT13 - Elitex (Lycra)	CT27 - Plug and Wear Nm10/3
CT14 - Shieldex 235/34x4 HC+E	CT28 - SilverSpun

Conductive pigment:	Dupont PE825 Silver Composite Conductor
TC pigments:	TC pigments Varioterm AC base 27°C - orange, red and black
Binder:	Gilaba vinyl acrylic binder
Finishing agents:	Aqueous emulsion vinyl acetate Aktilem 14000 and aqueous emulsion acrylic Aktilem 13002
Textile substrate:	50% cotton (CO) 50% polyester (PES) plain weave, 115g/m2 15,0 warp and 15,5 weft (tex), density of 41 ends/cm and 31 picks/cm

EQUIPMENT & METHODS

Samples were woven in an electronic Jacquard loom with 3456 hooks and a Bonas Series 200 Controller, with a plain weave structure. The substrate setup presents a 17 Tex PES texturized yarn with a density of 38 ends/cm. The weft comprises of 25,0 TEX CO (bleached) yarn and conductive threads, with a density of 11 picks/cm: 10 picks CO yarn plus 1 pick conductive thread. Weaving schematics of the samples developed are presented in 4.3.1 section.

Screen printing processes were conducted in a Zimmer Mini MDF R541 table. TC pigments were printed with the following settings: magnetic field level 3, application speed 50% of the graded dial and higher rod diameter parameter. A 12 mm metallic rod-squeegee and an open screen with 101,6 TPI were used. After drying, samples were thermo set in a laboratory oven at 150°C for 3 minutes.

Fig. 4.1 Materials, equipment and methods applied in the experimental work

Finishing agents were combined as follows: 15% Aktilem 14000, 15% Aktilem 13002 and 70% binder. The finishing layer was screen printed and thermo set with equal parameters as TC pigments.

The studies conducted with conductive pigment have tested different options of the screen printing table and will be described in 4.3.2 section. Thermo set parameters were 140°C during 10 minutes.

Color change activation through resistive heating was conducted with DC power supply Velleman LABPS3003SM.

To study samples' temperature during resistive heating activation, thermal images were recorded with a Testo 876 Infrared camera and analysed with Testo IRSoft software.

Images of samples screen printed with conductive pigments were also recorded with a magnifying glass Leica EZ4, 30x magnification.

Fig. 4.1 (continued)

4.3.1 Conductive Threads

This experimental section is aimed to integrate conductive threads in plain weave structures, to study their resistive heating behavior and respective activation of TC textiles color change.

Two electroconductive samples were woven. Sample A integrates 28 conductive threads, according to the schematic presented in Fig. 4.2. Each conductive thread accomplished two weft insertions of 15 cm, at a distance of 1 cm, totalizing 31 cm length plus the floating ends for connection with a power supply. The distance between each conductive thread type was also 1 cm. The conductive threads used are listed in Fig. 4.1, in respective order of weaving integration, from CT1 to CT28.

Sample B integrates four conductive threads, each accomplishing twelve weft insertions of 11 cm each, also at a distance of 1 cm, totalizing approximately 143 cm length. Figure 4.3 presents the schematic of the conductive threads weaving pattern and type.

After being woven, samples were screen printed (sample A was screen printed with 10% TC orange and sample B with 9% TC black and 6% TC red). Conductive threads number 5, 6 and 11 have a plastic insulated layer and they were tested in the thermo set conditions, previously to the definition of their integration in the substrate. The threads demonstrated stability to the conditions submitted.

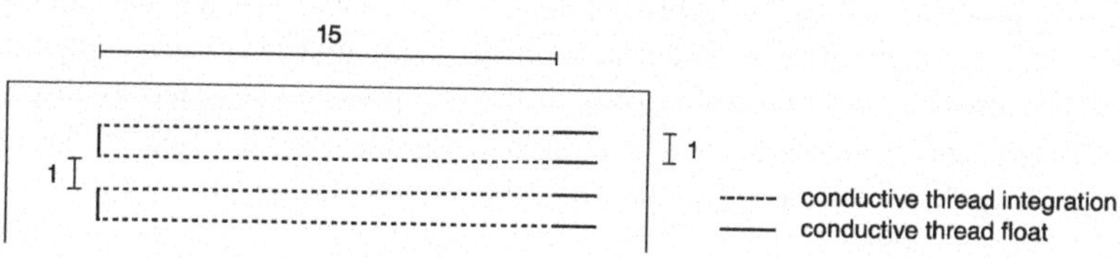

Fig. 4.2 Schematic of the conductive threads integration in sample A woven structure

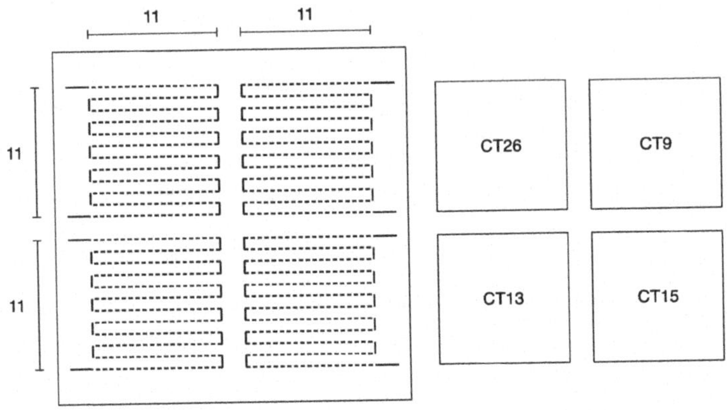

Fig. 4.3 Schematic of the conductive threads integration in sample B woven structure

Sample A experiments were conducted at a room temperature of approximately 22 °C, where the minimum voltage required for each conductive thread to heat up above 27 °C and thus start to change the textile color was tested and supplied for approximately 15 s. Electrical current attained was registered and a thermal image and photo were recorded with an IR camera. This process was repeated for voltage value increase, which was performed in a sequence of steps during 15 s each, until a maximum temperature of approximately 45 °C was detected through the IR camera. In sample B, a similar test was conducted at a room temperature of approximately 20 °C and each stage of power supply last 30 s. All samples were tested in horizontal positioning and placed above a Polyethylene (PE) board.

With the objective to work with folded structures and shape change behavior, electrical resistance of the conductive threads was measured with a multimeter, during handling of the sample from flat state to folded with a crease perpendicular to the conductive threads integration.

The set of conductive threads integrated in sample A have different visual and tactile characteristics, which is possible to perceive through the textile surface appearance and handling. Figure 4.4 presents sample A below 27 °C, with perspective image of sample top side (left); front view of the sample section from CT5 to CT11 (right, top) and backside (right, bottom).

The plastic insulated layer of CT5, 6 and 11 have affected the screen printed result with the lack of pigment in some areas. Other non-insulated threads that are as thick or more than the plastic insulated threads, have not presented issues during the printing process as they are more malleable. Conductive materials' thicknesses and colors also attained varying integration levels in the textile substrate appearance, being observable through the light color of the TC pigment selected as discrete or evident. Furthermore, some threads have also created a wrinkle effect in the textile, which was found to be more obvious after the thermo fixing process in particular with CT13, 16, 17 and 27.

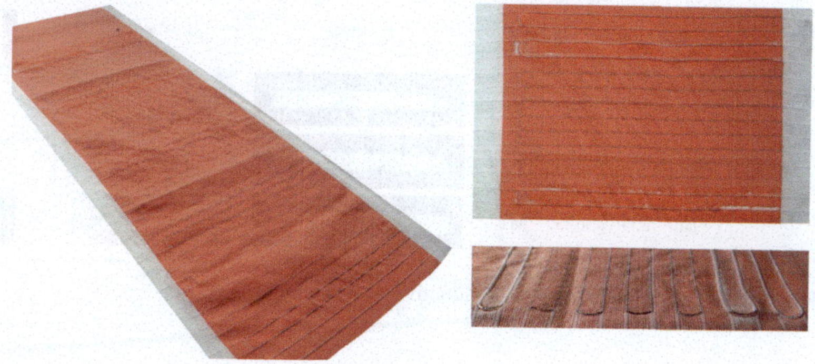

Fig. 4.4 Sample A: perspective of sample top side (left); front view from CT5 to CT11 (right, top) and perspective back side (right, bottom)

The tests conducted with sample A involved a set of relations between variables as identified:

- All conductive threads were integrated with the same pattern and length in the textile structure, as well as screen printed and thermo fixed parameters. Thus, besides the conductive threads' type, all the variables of the sample production were maintained constant.
- During the tests, room temperature, test setup, sample positioning and time between each power supply stage were also equal.
- The dependent variables were: conductive threads characteristics (materials, production process, electrical resistance, thickness, stiffness and color); power supplied at each stage (defined according to the start of color change until attaining approximately 45 °C) and electrical current in respect to the conductive thread electrical resistance; interaction of the conductive thread temperature(s) with the textile substrate and printed TC layer.

Considering the relations between variables, a framework was defined for the results documentation. The test conducted with each conductive thread was considered as a 'system', where the independent variable was voltage, affecting electrical current according to each thread electrical resistance, temperature and textile color change effect (dependent variables). Data was organized in respect to each conductive thread with the dependent results grouped in relation to each voltage stage. The analysis presented in this section includes specific data results according to the discussed findings. Complete documentation of the results is presented in Appendix A.

The results attained with CT1 are presented in Fig. 4.5, also displaying the conductive threads' electrical resistance calculated and the IR color scale defined. With 1.6 V power supply, CT1 attained 0.16 A, electrical current found required in the test setup to start to observe color change, as depicted in the first image of the photo column, recorded after 15 s of supplied power. The first IR image shows that at this stage,

CT1	V (V)	I (A)	R (Ω)	Tmax.(°C)
	1.60	0.16	10.00	34.2
	1.80	0.18	10.00	35.8
	2.00	0.19	10.53	38.6

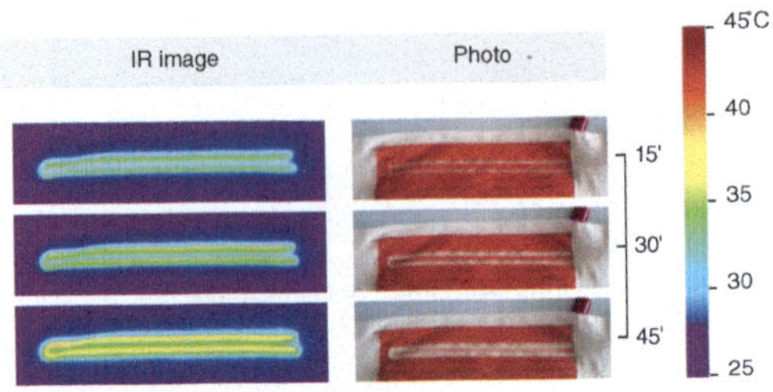

Fig. 4.5 Results of test conducted with conductive thread number 1 (CT1)

temperature in the conductive thread and nearby areas ranges in between approximately 30 and 34 °C. With this thread, each voltage variation stage was defined with a 0.2 V increase and the results show how electrical current and temperature in the thread varied and affected the textile heat transfer performing a pattern of color change from thin to thick stripes.

During the experiments conducted it was observed that aside each conductive thread required a specific electrical current value to attain a temperature above 27 °C, heat along the conductive threads' length presented different behaviors. In CT1 length, temperature appears to be similar, attaining color change stripes with similar thicknesses during power supply, whereas other threads attained distinct temperatures along their length. To analyze and discuss resistive heating behavior of each conductive thread, they were distinguished as homogeneous or heterogeneous, where in the last group, two main expressions were observed—irregular lines and traces. Figure 4.6 presents one example of each of these classes, with data results of each thread in the third voltage stage corresponding to 45 s of power supply.

As CT1, CT5 also exemplifies the homogeneous resistive heating in the thread length resulting in similar thicknesses along the color change lines. In CT4 and CT13, temperature along their lengths is heterogeneous, observed with the colors variation in IR images and color change effects in the photos. The results also demonstrate the two main sub-categories of the expressions observed with heterogeneous temperatures. In irregular lines (CT4), heat variation along the thread appears to maintain a small temperature variation along the thread length presenting some areas with

CT n°	V (V)	I (A)	R (Ω)	Tmax.(°C)
CT 5	3.00	0.25	12.00	42.0
CT 4	25.00	0.02	1250.00	41.6
CT 13	2.50	0.11	22.73	43.3

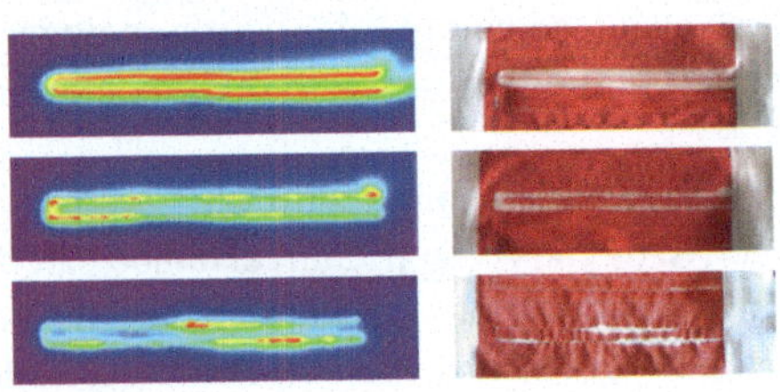

Fig. 4.6 Results attained with CT5, CT4 and CT13, in the 3rd stage of power supply defined (45″)

increased temperature and attaining a linear expression of color change with different thicknesses. In traces (CT13), the heat differences along the conductive thread have a larger interval and color change is only observed in some textile sections of the integrated conductive thread.

The irregular heat distribution in the threads can be interpreted as an incorrect behavior for resistive heating purposes. Indeed, in few cases increased attention was required to not overheat the textile, interrupting the power supply before the stipulated 15 s per stage. Alternatively and through a design perspective, the graphic effects analyzed demonstrate the possibility to create different color change expressions with conductive threads, which appear challenging to be produced by other means. Adding new languages to the chromic patterns, heterogeneous resistive heating of conductive threads can be explored in TC textile design, considering the possibility of controlling maximum temperatures in the thread or in applications where high temperature in the substrate is not an issue.

Furthermore, temperature variation along each thread length is not related to the absolute values of the conductive threads' electrical resistance and electrical current flow, rather it is also associated with the threads' materials and structure characteristics. Figure 4.7 presents the results of three conductive threads with different relations. CT15 and CT27 have the same electrical resistance and distinct heat behavior in the threads, as observed in the IR images and photos recorded during similar electrical current flow, attaining distinct expressions: traces and homogeneous, respectively. CT15 is an example where the temperature increased quickly at a specific point, exceeding 45 °C. Comparing CT26 and CT27, the inverse relation was observed, both presenting homogeneous color change lines but different electrical resistances.

In addition, small irregularities in the edge of color change stripes can also be observed with some conductive threads that present homogeneous temperature in their length. This effect occurred in threads that created a physical wrinkle in the textile substrate, affecting the heat transfer on the printed surface, as observed in CT16 (Fig. 4.8).

Thermal expansion in the textile areas between the threads' lines also differed. Comparing examples with similar temperatures in each conductive thread, some were able to perform a heat transfer above 27 °C in the textile areas between them with the

CT nº	V (V)	I (A)	R (Ω)	Tmax.(°C)	IR image	Photo
CT 15	6.00	0.04	150.00	49.2	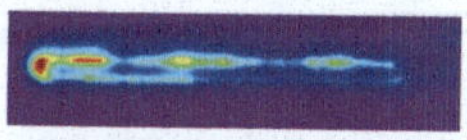	
CT 26	1.50	0.30	5.00	41.0	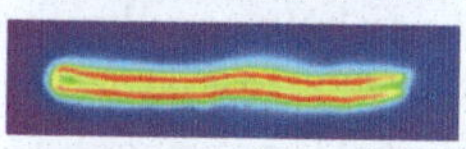	
CT 27	9.00	0.06	150.00	44.2	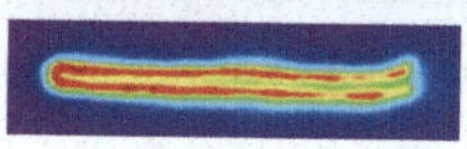	

Fig. 4.7 Results attained with CT15, 26 and 27, in the 3rd stage of power supply defined (45″)

CT n°	V (V)	I (A)	R (Ω)	Tmax.(°C)	IR image	Photo
CT 16	10.00	0.04	250.00	38.9	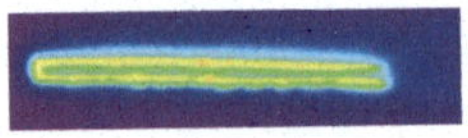	

Fig. 4.8 CT16 results in the 3rd stage of power supply defined (45″)

CT n°	V (V)	I (A)	R (Ω)	Tmax.(°C)	IR image	Photo
CT 9	3.00	0.16	18.75	39.5	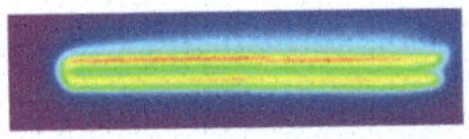	
CT 20	8. 00	0.06	133.33	39.5	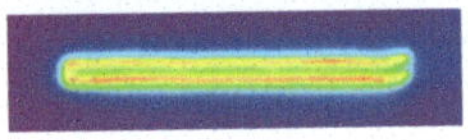	

Fig. 4.9 Results attained with CT9 and CT20, in the 3rd stage of power supply defined (45″)

time frame defined, while others did not. Figure 4.9 shows CT9 and CT20 results, where is possible to observe similar temperatures in each thread but different heating results in the textile. CT20 attained a complete color change in the area between their lines while CT9 still presented an area with the TC pigment in colorized state. This effect reflects the influence of the conductive threads materials and manufacturing processes in the heat transfer to the textile substrate.

Electrical resistance measurement during the sample folding attained different levels of values variation in each thread. The conductive threads integrated in the woven and printed sample that presented more stable results were CT1, 2, 7, 8, 9, 19, 21, 23, 24, 25 and 26.

To further explore the linear woven pattern of conductive threads and color change in larger areas of the textile, sample B was developed with four conductive threads, one for each category defined: homogeneous with higher and less thermal expansion (CT26 and CT 9, respectively), heterogeneous with irregular lines and traces (CT15 and CT13, respectively). The sample was screen printed with a darker TC colour, to observe the changes through increased colour luminosity difference between temperature variation. Data documentation followed the previous setup, presenting the results of each conductive thread at the second voltage stage defined, corresponding to one minute of power supply (Fig. 4.10).

CT26 and CT9 were selected to observe homogeneous heating with different thermal expansions, in each square area of their integration. The results attained have shown differences between the threads behaviour than in sample A, where CT26 heated completely the textile area in between their lines with a lower electrical current (0.3 A) while CT9 presented a similar effect. The heterogeneous heating along CT13 length displayed variations of colour change stripes thickness, but also required higher electrical current to heat up above 27 °C in comparison with sample A. In CT15, the last stage of power supply was 30 V (the limit of the equipment

CT n°	V (V)	I (A)	R (Ω)	Tmax.(°C)	IR image	Photo
CT9	20.00	0.23	86.96	42.9	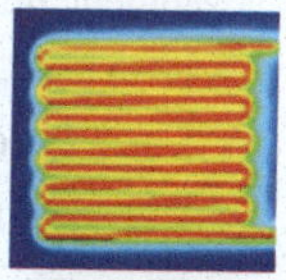	
CT13	11.00	0. 29	37.93	42.0	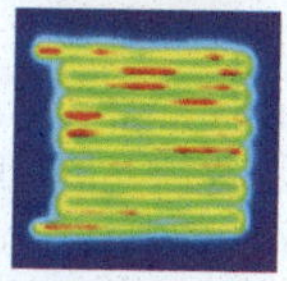	
CT15	30.00	0.06	500.00	39.8	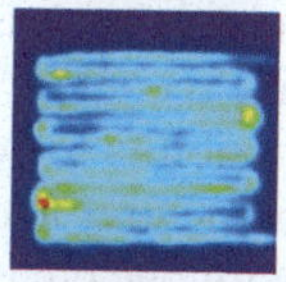	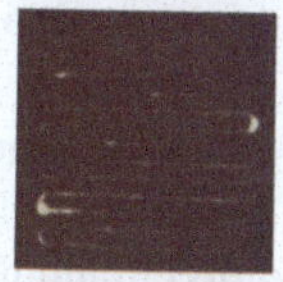
CT26	10.00	0.42	23.81	44.2	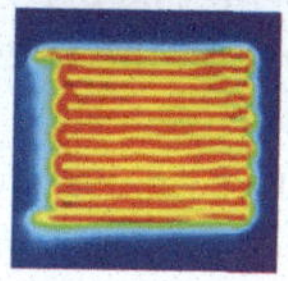	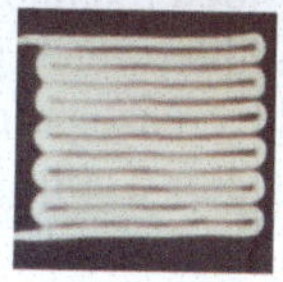

Fig. 4.10 Sample B results with CT9, 13, 15 and 26 in the 2nd stage of power supply defined (1′)

used) attaining a maximum temperature of 39.8 °C and colour change effect with colourless traces of varyied thicknesses.

This test was conducted at a slighly lower room temperature than sample A tests, while the time of power supply was increased. The results attained highlighted that these variables strongly affect textile temperature and colour change effect, having delineated further studies to systematically analyse their influence in textile thermo-responsive behaviour, which will be presented in Chap. 7.

4.3.2 *Conductive Pigments*

This experimental section is aimed to screen print conductive pigments in textiles, to study their resistive heating behavior and respective activation of TC textiles color change.

A silver-based conductive pigment was used to analyze the effect of screen printing parameters in the electrical resistance of a printed traces' set. Conductive traces of 10.0 × 0.2 cm were screen printed with variation of the follow parameters: magnetic field applied by the screen printing table; number of screen printed layers and textile substrate without and with treatment of a finishing layer. Table 4.1 describes the samples developed. Samples' screen printing sequence was: finishing layer (for samples treated); conductive pigment; and TC paste.

Table 4.1 Description of conductive pigment samples C and F

	Magnetic field 3			Magnetic field 6		
	1 layer	2 layers	3 layers	1 layer	2 layers	3 layers
Substrate	C1	C2	C3	C4	C5	C6
Treated substrate	D1	D2	D3	D4	D5	D6

For a comparative analysis of samples electrical resistance and resistive heating behavior, a test was conducted with a similar setup as described in Sect. 4.3.1 for sample A, at a room temperature of approximately 20 °C. For each conductive trace of silver-based pigment, 1.5 V was supplied for 15 s, the current attained was registered and a thermal image recorded.

To study the possibility to develop a conductive printed circuitry to activate color change in folded structures and shape changing textiles, each conductive trace was folded manually with one crease in the middle. For an analysis of the folding effect in the printed pigment, the tests previously conducted were repeated and images of the crease point were recorded with a magnifying glass.

For a comparative analysis of the results, Fig. 4.11 presents the data arranged in a framework, according to samples' relation between textile substrate and treated substrate, table magnetic field (mf) and screen printing layers number. Conductive traces screen printed in the treated substrate attained slightly lower electrical resistance values than the correspondent traces developed in the untreated textile. The increase number of printed layers also resulted in samples' electrical resistance decrease, with larger value variation in substrates without finishing layer. Conductive traces printed with the magnetic field 6 (mf 6) presented higher resistance than with mf3, also with larger differences in samples with the untreated substrate.

Resistive heating behavior depicted by thermal images shows that heat in each conductive trace is not homogeneous, higher temperatures were attained in the edges of each trace length, observed in most of the samples with two red lines and yellow in the middle. The effect of the printing parameters in the temperatures attained appeared

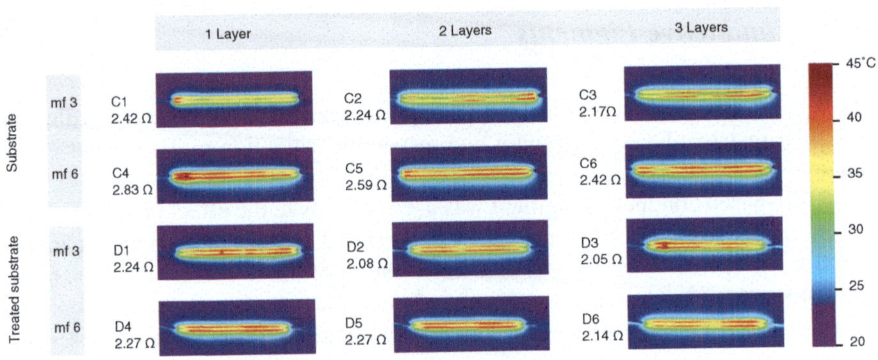

Fig. 4.11 Electrical resistance and IR images of samples C and D, with 1.5 V power supply

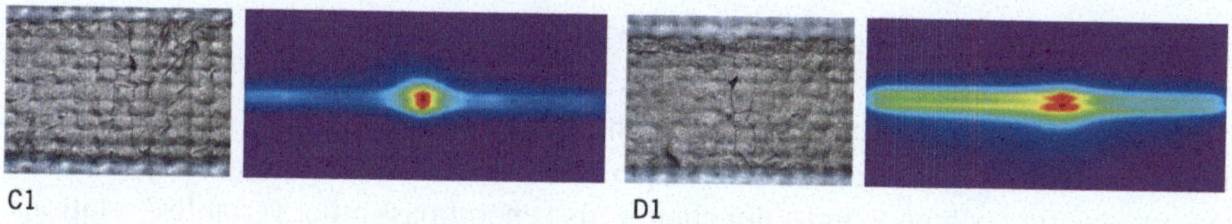

Fig. 4.12 Magnifying glass photo and IR image of C1 and D1, after being folded

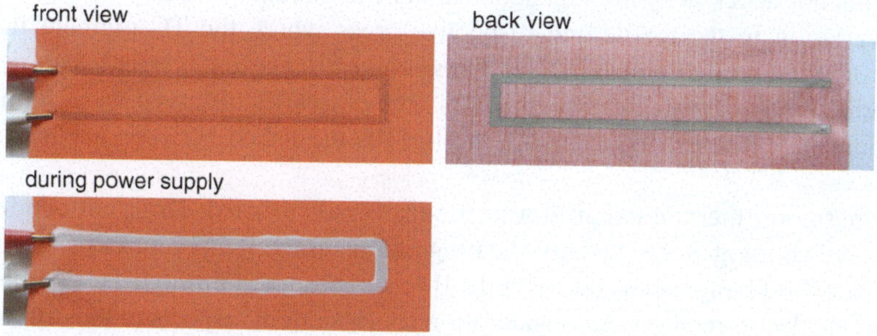

Fig. 4.13 Conductive pigment sample front and back view before power supply (top left and top right, respectively) and during power supply (bottom)

subtle. The more representative relation was observed in regards to conductive traces screen printed with mf 6 presenting a slight increase of temperatures attained in comparison to mf 3.

After the presented analysis, the folding effect on conductive traces was studied. The magnifying glass record (30× magnification) has shown that all conductive traces crackled in the crease line, being more visible for samples with treated substrate. Resistive heating tests demonstrated an overheating effect in the folded areas, with a more concentrated heat spot in samples with untreated substrate. Figure 4.12 presents C1 and D1 results.

To illustrate color change with the studied conductive pigment, Fig. 4.13 presents a sample developed with the same printing parameters of the conductive trace C6 in the textile backside and 10% TC orange in the front. Printed on the samples' back, the conductive pattern is visually perceived through the TC pigment layer before activation (front view image). With power supplied, the sample changed color in the area of the conductive pigment pattern, whereas thermal expansion in the textile was reduced considering a maximum temperature of 45 °C during activation.

4.4 Conclusions

The integration of conductive materials in textile substrates enables the creation of soft structures with electrical properties. In this research section, electroconductive textiles were developed to study integration processes of conductive threads and pigments in textiles and their resistive heating properties towards TC pigments color change activation.

Conductive materials electrical resistance and electrical current required to heat up above the TC pigment activation temperature are important parameters. However, resistive heating behavior and color change also encompass other variables' relations.

Integration pattern of conductive threads in the woven structure defines the possible areas of textile heat transfer. The results attained have demonstrated that resistive heating is not necessarily homogeneous along the threads length, affecting temperature variation in the textile and respective areas where the TC pigment transits from colorized to decolorized state. Thus, color change expressions also depend on conductive threads expressive qualities. The outcomes observed were defined as homogeneous or heterogeneous with the last distinguished between irregular lines and traces.

Furthermore, thermal expansion in the textile also varied. Different conductive threads with homogeneous resistive heating and at similar temperatures have attained different color change areas between the threads, observed through TC complete or incomplete decolorization. A consideration for this effect respects to the differences between the conductive threads fabrication, for example a textile metal spun thread or a textile metal plated, resulting in different contact areas between the conductive materials and the textile hold structure consequently affecting textile heat transfer.

The study conducted with silver-based pigment observed that printing parameters affect the electrical properties of the conductive printed pattern. However, the differences attained were not significant in the resistive heating outcomes. Taking into consideration the results attained with textile substrate handled and processes developed, the conductive pigment studied have presented great potential to explore color change patterns with textiles developed by printing processes, but did not showed satisfactory resistive heating behavior for the development of textile folded structures.

Towards the development of study samples and prototypes to further research textile dynamic behavior, a conductive material was selected. Considering the results attained, conductive threads presented greater properties to be integrated in color and shape change textiles due to electrical resistance stability during textile folding states, particularly with CT1, 2, 7, 8, 9, 19, 21, 23, 24, 25 and 26. Additionally, their integration in a plain weave structure also allows the folding of textile piled up layers, without requiring electrical insulation.

The conductive thread selection also took into consideration the electrical resistance level, heating behavior, possibility of programmed integration in the weaving process (not manually), tactile and visual characteristics.

Compared to the substrate textile threads used, CT2 is very thick and affected the textile sample tactility and folded morphology, reason why CT24 and 26 were also not selected, given their stiffness. CT23 and 25 can be broken manually and thus it was considered that their low tensile strength could present issues to support the tensions exposed during the programmed weaving process.

Furthermore, electrical resistance and resistive heating behavior determined the conductive thread selection in respect to the objectives of the research prototypes scale, morphologies and activation. With the aim to activate complete areas of color change, homogeneous heat along the threads' length was selected, as it enables similar heat transfer in the textile areas in between the conductive threads. CT21 presented heterogeneous heating and thus it was not considered.

The textile activation was also defined to be performed through different conductive series circuits, each being integrated in larger areas than, for example, sample B (11 × 11 cm integration area of each conductive thread). This scale estimation has considered the size decrease within folded origami tessellations and the objective to integrate conductive threads in the overall textile area. Besides each material type, each circuit electrical resistance depends on the conductive thread length used and as larger the circuit as higher the voltage required to attain a specific electrical current (V = I × R). It was then defined to choose a low resistance thread and the selection fallen into CT7 (High Flex Silver), also due to their lighter color, compared to CT8, presenting a more discrete appearance in the textile printed surface. A woven test was conducted in the loom Jacquard Vamatex with 41.2 Tex CO warp and 14.7 Tex PES weft, and the programmed integration of the CT7 in the woven substrate was successfully verified. Considering this research aims, CT7—Karl Grimm High Flex 3981 7 × 1 Silver—was selected due to their resistive heating properties, processability, soft tactility and light color.

Regarding data documentation, the intrinsic properties of the conductive material studied comprise technical and expressive qualities, which were analyzed as interconnected relations with the textile chromic behavior. The way the results were framed aimed to depict technical and design findings as well as their relations, exploring a vocabulary that connects abstract with visual languages towards a comprehensive documentation within a technical and design approach of materials resistive heating and color change behavior.

This method associated with the frameworks organization in the conductive materials catalogue, built a meaningful tool to research, analyze, compare and communicate conductive materials and textile behavior. This documentation process was implemented in further experimental studies, conducted during the *Dynamic Light Filters* design research.

References

Baurley S (2005) Interaction design in smart textiles clothing and applications. In: Tao X (ed) Wearable electronics and photonics. Woodhead Publishing, Cambridge

Bosowski P, Hoerr M, Mecnika V, Gries T, Jockenhövel S (2015) Design and manufacture of textile-based sensors. In: Dias T (ed) Electronic textiles. Elsevier Science & Technology, Cambridge

Chatterjee A, Maity S (2018) Electroconductive textiles. In: Shahid-Ul-Islam, Butola BS (eds) Advanced textile engineering materials. Wiley and Scrinever Publishing, Beverly

Collection (n.d.) Merriam-Webster's online dictionary [Online]. https://en.oxforddictionaries.com/definition/collection. Accessed 18 Feb 2017

Cork C (2015) Conductive fibres for electronic textiles: an overview. In: Dias T (ed) Electronic textiles. Elsevier Science & Technology, Cambridge

Dias T, Ratnayake A (2015) Integration of micro-electronics with yarns for smart textiles. In: Dias T (ed) Electronic textiles. Elsevier Science & Technology, Cambridge

Eichhoff J, Hehl A, Jockenhoevel S, Gries T (2013) Textile fabrication technologies for embedding electronic functions into fibres, yarns and fabrics. In: Kirstein T (ed) Multidisciplinary know-how for smart textiles developers. Woodhead Publishing, Oxford

Espace GZ (n.d.) Thermo-regulating systems [Online]. http://www.gradozero.eu/gzenew/index.php?pg=thermaltech&lang=en: Grado Zero Espace. Accessed 2 Nov 2015

Gibilisco S (2001) The illustrated dictionary of electronics. McGraw-Hill, New York

Gordon P, Russel T, Steve B, John T (2014) An investigation into the durability of screen-printed conductive tracks on textiles. Meas Sci Technol 25

Guo L, Bashir T, Bresky E, Persson N (2016) Electroconductive textiles and textile-based electromechanical sensors—integration in as an approach for smart textiles. In: Koncar V (ed) Smart textiles and their applications. Woodhead Publishing, Amsterdam

Hu E, Kaynak A, Li Y (2005) Development of a cooling fabric from conducting polymer coated fibres: proof of concept. Synth Metals 150(2). https://doi.org/10.1016/j.synthmet.2005.01.018

Jin H, Matsuhisa N, Lee S, Abbas M, Yokota T, Someya T (2017) Enhancing the performance of stretchable conductors for E-textiles by controlled ink permeation. Adv Mater 29:1605848-n/a. https://doi.org/10.1002/adma.201605848

Kazani I (2012) Study of screen-printed electroconductive textile materials. PhD thesis, University of Ghent

Kirstein T (2013) The future of smart-textiles development: new enabling technologies, commercialization and market trends. In: Kirstein T (ed) Multidisciplinary know-how for smart textiles developers. Woodhead Publishing, Oxford

Koncar V (2016) Introduction to smart textiles and their applications. In: Koncar V (ed) smart textiles and their applications. Woodhead Publishing, Amsterdam

Kumar L, Vigneswaran C (2016) Electronics in textiles and clothing: design, products and applications. CRC Press, Taylor & Francis Group, Boca Raton, FL

Langenhove LV (2015) Smart textiles: past, present, and future. In: Tao X (ed) Handbook of smart textiles. Springer, Singapore

Li Q, Tao X (2015) Fabric substrates and interconnectors for three-dimensional surfaces. In: Tao X (ed) Handbook of smart textiles. Springer, Singapore

Lin S (2015) Fiber-based wearable electronic circuits and systems. In: Tao X (ed) Handbook of smart textiles. Springer, Singapore

Locher I (2013) Joining technologies for smart textiles. In: Kirstein T (ed) Multidisciplinary know-how for smart textiles developers. Woodhead Publishing, Oxford

Matsuhisa N, Kaltenbrunner M, Yokota T, Jinno H, Kuribara K, Sekitani T, Someya T (2015) Printable elastic conductors with a high conductivity for electronic textile applications. Nat Commun 6:7461. https://doi.org/10.1038/ncomms8461

Mbise E, Dias T, Hurley W (2015) Design and manufacture of heated textiles. In: Dias T (ed) Electronic textiles. Elsevier Science & Technology, Cambridge

Mecnika V, Scheulen K, Anderson C, Hörr M, Breckenfelder C (2015) Joining technologies for electronic textiles. In: Dias T (ed) Electronic textiles. Elsevier Science & Technology, Cambridge

Mehmann A, Varga M, Tröster G (2017) Reversible contacting for smart textiles. In: Schneegass S, Amft O (eds) Smart textiles: fundamentals, design, and interaction. Springer International Publishing, Cham

Miao M (2015) Carbon nanotube yarns for electronic textiles. In: Dias T (ed) Electronic textiles. Elsevier Science & Technology, Cambridge

Montazer M, Harifi T (2018) Nanofinishing of textile materials. Woodhead Publishing, Cambridge

Nuno (n.d.) About Nuno [Online]. https://www.nuno.com/en/about/: Nuno Corporation. Accessed 20 Oct 2015

Pailes-Friedman R (2016) Smart textiles for designers: inventing the future of fabrics. Laurence King Publishing, London

Qu H, Skorobogatiy M (2015) Conductive polymer yarns for electronic textiles. In: Dias T (ed) Electronic textiles. Elsevier Science & Technology, Cambridge

Rakshit A, Hira M (2014) Electrically conductive fibre substrates. Int J Fiber Text Res 4:44–48

Schneegass S, Amft O (2017) Introduction to smart textiles. In: Schneegass S, Amft O (eds) Smart textiles: fundamentals, design, and interaction. Springer International Publishing, Cham

Schwarz A, Langenhove LV (2013) Types and processing of electro-conductive and semiconducting materials for smart textiles. In: Kirstein T (ed) Multidisciplinary know-how for smart textiles developers. Woodhead Publishing, Oxford

Shim E (2010) Coating and laminating processes and techniques for textiles. In: Smith WC (ed) Smart textile coatings and laminates. Woodhead Publishing, Cambridge

Stamatellos G (2007) Computer ethics: a global perspective. Jones and Bartlett Publishers, USA

Stoppa M, Chiolerio A (2014) wearable electronics and smart textiles: a critical review. Sensors 14:11957. https://doi.org/10.3390/s140711957

Storey N (2009) Electronics: a systems approach. Pearson Education Limited, Harlow

Suganuma K (2014) Introduction to printed electronics. Springer, New York

Tao X, Koncar V (2016) Textile electronic circuits based on organic fibrous transistors. In: Koncar V (ed) Smart textiles and their applications. Woodhead Publishing, Amsterdam

Tong J, Li L (2015) Thermal regulation of electrically conducting fabrics. In: Tao X (ed) Handbook of smart textiles. Springer, Singapore

Toprakci H, Ghosh T (2015) Textile sensors. In: Tao X (ed) Handbook of smart textiles. Springer, Singapore

Torah R, Wei Y, Li, Y, Yang K, Beeby S, Tudor J (2015) Printed textile-based electronic devices. In: Tao X (ed) Handbook of smart textiles. Springer, Singapore

Varga M (2017) Electronics integration. In: Schneegass S, Amft O (eds) Smart textiles: fundamentals, design, and interaction. Springer International Publishing, Cham

Veja P (2014) An investigation of integrated woven electronic textiles (e–textiles) via design led processes. PhD thesis, Brunel University

Wagner M (2013) Automotive applications of smart textiles. In: Kirstein T (ed) Multidisciplinary know-how for smart textiles developers. Woodhead Publishing, Oxford

Wang L, Wang X, Lin T (2010) Conductive coatings for textiles. In: Smith WC (ed) Smart textile coatings and laminates. Woodhead Publishing, Cambridge

Wang Z, Volakis J, Kiourti A (2015) Embroidered antennas for communication systems. In: Dias T (ed) Electronic textiles. Elsevier Science & Technology, Cambridge

Warmx (n.d.) Smart textiles—product development around textile current circuits [Online]. http://www.warmx.de/index.php/industry-and-research.html. Accessed 25 Oct 2015

Wilson J, Mather R (2015) Photovoltaic energy harvesting for intelligent textiles. In: Dias T (ed) Electronic textiles. Elsevier Science & Technology, Cambridge

Xue P, Tao X, Leung M-Y, Zhang H (2005) Electromechanical properties of conductive fibres, yarns and fabrics. In: Tao X (ed) Wearable electronics and photonics. Woodhead Publishing, Cambridge

Yang K, Torah R, Wei Y, Beeby S, Tudor J (2013) Waterproof and durable screen printed silver conductive tracks on textiles. Text Res J 83:2023–2031. https://doi.org/10.1177/0040517513490063

Yokus MA, Foote R, Jur JS (2016) Printed stretchable interconnects for smart garments: design, fabrication, and characterization. IEEE Sens J 16:7967–7976. https://doi.org/10.1109/JSEN.2016.2605071

Zeng W, Shu L, Li Q, Chen S, Wang F, Tao X-M (2014) Fiber-based wearable electronics: a review of materials, fabrication, devices, and applications. Adv Mater 26:5310–5336. https://doi.org/10.1002/adma.201400633

Part II
Dynamic Form in Textiles

Chapter 5
Dynamic Form

The possibility of conferring dynamic form in textile structures creates an emergence of innovative properties and interactions for smart textiles. Thermo-responsive SMAs are metal compounds capable of changing from a temporary to a pre-programmed shape, upon a heat stimulus. By converting thermal into mechanical energy, they enable the transfer of the shape change effect to a host textile structure.

Considering the Dynamic Light Filters concept, textile behavior aims to change the light that passes through them. As different amounts of textile layers absorb different luminosities, the research on shape memory textiles aimed to create changes within layer number variation, according to predefined geometric morphologies. These requirements were studied through origami techniques.

This chapter introduces base concepts associated with form, shape changing interfaces and origami structures. A review on SMMs and SMAs properties and behavior is presented, followed by an analysis of the process to memorize the alloys shape, developing technologies for their integration in textile substrates and applications.

5.1 Form and Origami Surfaces

Shape memory textiles behavior comprehends a material system that performs physical transformations. Regarding temporal and spatial relationships, shape change can be visually perceived as movement (Gordon 2004). Along speed and direction parameters, motion perception also includes the ability to recognize a shape in different positions or seen from different viewpoints, called viewpoint invariance (Goldstein 2010). Thus, during shape change, two configurations are perceived as different positions or orientations of the same shape.

I. D. Cabral et al., *Dynamic Light Filters*, Springer Series in Design and Innovation 3, https://doi.org/10.1007/978-3-030-39529-2_5

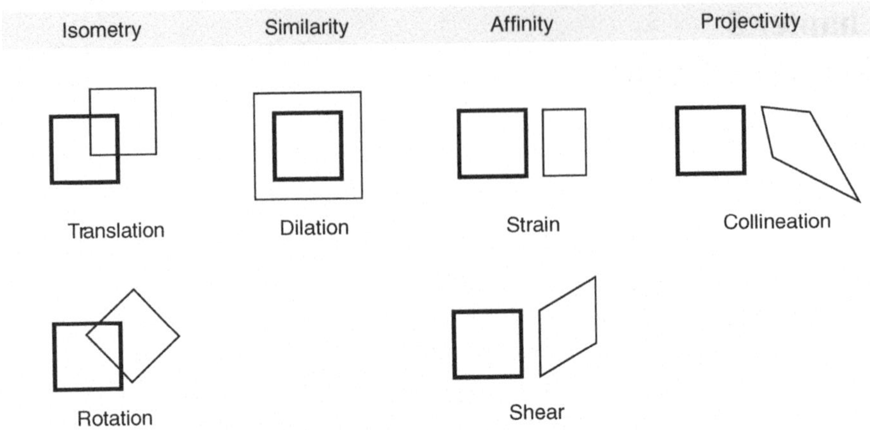

Fig. 5.1 Klein transformations (Nefs 2008)

In nineteenth century, the German mathematician Felix Klein defined a group of shape transformations perception, according to invariant geometric properties (Nefs 2008). Figure 5.1 illustrates the four groups: isometry, similarity, affinity and projectivity. The original images are depicted with a thicker line than the transformed images and they are represented in 2D but can also be reported for three-dimensional shapes.

Isometry is the change in position of equal shapes, they maintain the angles and distance of their geometry and vary position in space according to translation and/or rotation processes. In similarity transformation the points of an object maintain their angles and distance ratio with a point in the field, as can be observed in dilation. All isometries are also similarities and shape perception within this transformation allows us to recognize an object, for example, independently of its size or depth in the visual field. Affinity is the transformation that preserves parallel relations while position and distance ratios can vary. This transformation includes strain effect, where one dimension of the object can be increased or decreased and shear effect where parallelism is maintained within displacement in one dimension as a function of a second dimension, for example when a shape is distorted to an angle perspective. Finally, projectivity is the transformation where all collinear points (points that lie in a single straight line), maintain collinearity in the transformed object (Nefs 2008; Sibley 2015).

Furthermore, the visual system also attempts to organize what is perceived in order to make sense of the world. Based on the principle that the whole is different from the sum of the parts, Gestalt psychologists have proposed laws of perceptual organization (May 2007). The law of pragnanz defends that stimulus patterns of grouped elements are perceived and interpreted as simple as possible, meaning that in the presence of complex shapes, the visual system tries to organize them in the simpler way (Goldstein 2010). This law is central in the Gestalt psychology and based on it, several principles have been outlined, which can be relevant to be familiar

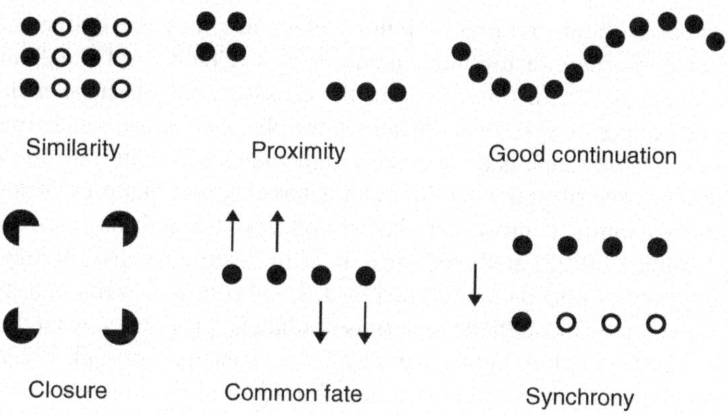

Fig. 5.2 Gestalt laws: similarity, proximity, good continuation, closure, common fate and synchrony

with, when working with shape memory textiles, such as: similarity, proximity, good continuation, closure, common fate and synchrony. Figure 5.2 presents an example for each of the aforementioned principles, also called laws.

Similarity law refers to the tendency of elements with similar characteristics appear to be grouped. In the illustrated example, dots of the same color can be perceived as vertical columns. Diverse characteristics can fulfil this effect such as shape and texture. The law of proximity refers to objects that are closer also appearing to be grouped, independent of their shapes. Good continuation refers to the tendency of perceiving elements that show smooth connections as a unit. Thus, shapes rearranged in lines or curves are more likely to be grouped. Closure means that it is possible to perceive complete figures, despite some parts are incomplete. The visual system is able to fill the gap of the missing information. Common fate law refers to objects that move in the same direction appearing to be grouped. Events that occur at the same time also can be perceived as belonging together and that is the law of synchronism (Gordon 2004; May 2007; Nefs 2008).

Gestalt principles are considered important in the visual perception field, for an understanding of how the visual system tends to organize what is observed. Nevertheless, the observation time can also play a crucial parameter of how images are perceived and usually longer time falls in a more complex and detailed analysis. Additionally, several visual illusions are also related with Gestalt laws (Nefs 2008).

In context of shape change textiles, the described principles highlight the possibilities of shapes being perceived as a unit or pattern according to their similarities, proximity and rearrangement; the perception of a complete shape being attained within incomplete images; and the activation of shape change across equal direction and/or time of event, being able to group individual elements.

The change in shape of physical interfaces can also be characterized according to different variables. Rasmussen et al. (2012) suggest eight types of changes that can occur individually or grouped: orientation, form, volume, texture, viscosity, spatiality, adding or subtracting and permeability. Regarding changes from one form to another without dividing or joining elements, they are defined in mathematics as homeomorphic or topological equivalents (Coelho and Zigelbaum 2011; Rasmussen et al. 2012). In this research, the development of shape memory textiles addresses material system with homeomorphic behavior, which trained and programmed mutable shape is triggered through changes in materials properties.

With the objective of understanding characteristics and qualities of movement, several authors examined movement-based concepts through an interdisciplinary approach. Vaughan (1997) analyzed movement in theatre and psychology, namely through the theory of affordances coined by J. J. Gibson, which refers to the action possibilities perceived through the materials or objects' properties, what they afford. Young et al. (2005) explored choreography based-concepts through Laban theory, a method developed for movement notation. Young used the metaphor of language considering the element of movement as the alphabet, which he defines as path, volume, velocity and direction to discuss sequence and qualities of movements as the grammar and syntax.

Rasmussen et al. (2012) discussed movement through a critical review of shape-changing interfaces, distinguishing kinetic parameters as the physical specifications of movement, based on Youngs' alphabet, and defined association and adjectives as the expressive parameters of how movement is perceived. He described expressive association is respect to the changes being perceived either mechanical or organic and the expressive adjectives as the traits and qualities assigned to the changes.

The interrelation between shape, materials and dynamic behavior is transversal to diverse domains focused in shape-changing structures. Hensel and Menges (2006) reflect upon self-organization examples in the natural world to study how form emerges and continually differentiates, transforms and performs. Through digital fabrication and performance analysis methods, they design new approaches on complex adaptive systems for architecture (Hensel et al. 2006). Concerning soft mechanic systems, Coelho and Zigelbaum (2011) focus on topological, textural and permeable transformations with SMAs, to analyze emergent possibilities for shape-changing applications.

Developing shape memory textiles with the ability to perform predefined geometric configurations also demands a study of the textile physical form or shape—morphology (Hensel and Menges 2006). Textile substrates can be manipulated or shaped to exhibit different textures or structures, where diverse techniques can be applied such as folding, smocking, 3D shibori and finishing processes (Richards 2012; Wada 2012; Singer 2013). In this study, the solution for the requirement of geometric rigor and generation of morphologies with different number of layers was developed through origami techniques.

Origami is an ancient Japanese art of folding paper that in the last decades has been capturing increased interest among the arts, design, engineering and science (Hartl et al. 2014; Li et al. 2019). Across diverse materials, origami techniques present new possibilities to fabricate, assemble, store and morph structures, concepts explored in a wide range of origami-based applications, from deployable space structures to DNA technologies at nano-scale, medical devices, collapsible shelters, robotics, fashion, packaging, among others (Miura 1994; Edwards and Yan 2014; Onal et al. 2015; Peraza Hernandez et al. 2019).

Surface effects, haptic qualities, sculptural and functional aims have been approached within origami textiles, where besides traditional methods to create origami models by manual experiments, there are also software applications which can generate crease patterns such as Origamizer, TreeMaker and Oripa (Tachi 2010; Lang 2012; Mitani 2011). Issey Miyake and his Reality Lab team developed—132 5. Issey Miyake—a label of "a piece of cloth", based on origami concepts and software development which explores how textile structures can extend from a flat foldable piece to a three dimensional garment or accessory (Shimizu 2016).

Active origami is an emergent research topic focused on structures with self-folding capabilities. Shape change is based on the use of active materials, which are able to convert an energy form such as thermal, chemical or electromagnetic, into mechanical work, thus enabling the creation of dynamic morphological performance without external manipulation (Peraza Hernandez et al. 2019).

Regarding the combination of active materials and origami, Peraza-Hernandez et al. (2014) highlight as critical design parameters the actuation strain, actuation stress and the necessity to balance the actuators characteristics within the structures' scale. Some examples of self-folding structures include: Programmable Matter, where a triangular tiled sheet was combined with SMA foil actuators and flexible electronics to create a material that performed a self-folding boat and airplane (Hawkes et al. 2010; Paik et al. 2011); a laminated composite with two outer layers of SMA film, able to form three-dimensional structures (Hartl et al. 2012); and paper transformations where SMAs wires were taped, sewed or attached with snaps (Koizumi et al. 2010; Saul et al. 2010; Qi and Buechley 2012).

Research in this area with other active materials include photo-origami through the use of a light-sensitive polymer film (Ryu et al. 2012), self-folding graphene paper with graphene oxide (Mu et al. 2015) and 3D printing with shape memory polymers or hydrogel, termed as 4D printing considering time as the fourth dimension (Ge et al. 2014; Zhang et al. 2019). This concept was extended to 4D textiles, defining 3D printed textile structures that can change form over time upon an external stimulus (Schmelzeisen et al. 2018).

In this research, dynamic behavior of origami-inspired textiles was studied by means of SMAs wires integration in woven substrates and aimed to create transformations between predefined geometric shapes, to attain layer variation and affect light transmittance. Thus, textile morphology seeks for structural and expressive possibilities, taking into consideration shape, material system and defined behavior as connected elements.

To illustrate the origami folds, a crease pattern or origami diagram is implemented. Crease patterns consist of drawing mountain and valley folded lines necessary to depict the origami creases (Jackson 2011). They can be represented in different ways. In this study, a black continuous line for mountain fold and red dashed line for valley fold were applied.

There are origami base folds that can be combined to develop more complex models (DiLeonardo-Parker 2016). Origami tessellation is based on repetition of common base patterns of folded pleats and twists, creating intricate patterns. These origamis can display a property of flat foldability, which means the origami model is flat after being folded (Gjerde 2009; Lang 2018). Surfaces folded through tessellation techniques create areas with different amount of layers.

In *shadowfolds*, Rutzky and Palmer (2011) combined the concept of pleat and translucency for origami tessellations developed in textiles. They highlighted the expressive quality that these models may exhibit such as the light shining through and the revelation of the folds' depth. In this research, the translucency of textile origamis is studied in a dynamic perspective, in which the interaction of the shape memory textile with light will design dynamic light for expressive and functional purposes.

5.2 Shape Memory Materials

Shape Memory Materials are a class of stimuli-responsive materials with the ability to change from a temporary to a memorized shape (Otsuka and Wayman 1998). Their unique behavior and properties hold great potential for applications where shape change across sensing and actuation capabilities are required and have been gaining increased attention in textile design and engineering domains (Sun et al. 2012; Lexcellent 2013).

The shape change performance of SMMs lies in a phase transformation induced by an external stimulus, after having been deformed to a temporary shape, they respond to an input by returning to a pre-programmed physical geometry. Thermal energy is the most common SMMs' sensing stimulus, meaning that they are activated at a critical temperature, also known as reverse transformation temperature (Otsuka and Kakeshita 2002). Moreover, there are also SMMs that respond to stress, magnetic or electric field, pH-value, UV light and water (Huang et al. 2005; Lendlein et al. 2005; Yeduru 2013).

The ability to perform shape change behavior has been found in certain alloys, polymers, gels and ceramics which is called Shape Memory Effect (SME) (Tao 2001; Honkala 2006). The materials' types considered more suitable to be integrated in textile substrates are the Shape Memory Alloys (SMAs) and Shape Memory Polymers (SMPs), in particular in wire and monofilament form. Numerous researchers have examined and characterized their properties and both materials present diverse advantages and limitations to take into consideration in each application requirements. SMPs are reported as having high shape deformability and recoverability, also being more economic and flexible than SMAs. However, SMAs offer superior mechanical and recovery strengths as well as time response (Hu 2007; Dyer 2010; Huang et al. 2010; Hu and Meng 2011; Sun et al. 2012; Jani et al. 2014; Gök et al. 2015).

In this research, higher actuation forces are a crucial parameter considering the objective to integrate SMMs in textile substrates to indirectly transfer the SME to the host structure, within predefined geometric morphologies. Therefore, the materials chosen to study were the SMAs.

Thermo-responsive SMAs are metal compounds capable of converting thermal energy into mechanical energy, commonly generating motion and force. As illustrated in Fig. 5.3, their behavior is based on two reversible phase transitions named: Austenite (A) at high temperature phase, also called parent phase; and Martensite (M) at low temperature phase, each one presenting different crystal structures and consequently different properties (Honkala 2006). When a SMA is deformed at M phase and temperature increases, it begins to transit to A phase at austenitic start temperature (As), completing the transformation at austenitic finish temperature (Af). There is a transformation of the alloy crystal structure that becomes cubic, after which the alloy returns to its memorized shape. By decreasing the temperature, the reverse transformation starts to occur at martensitic start temperature (Ms) and the process is concluded at martensitic finish temperature (Mf). There is no shape change due to this transformation. When Mf temperature is attained, the alloy is ductile and can be deformed. In this phase, if a shape transformation by external forces occurs, the alloy passes from twinned martensite to detwinned martensitic state (Humbeeck 2009; Adiguzel 2015; Rao et al. 2015).

The process described is known as the one-way memory effect, the alloy is able to memorize the shape that recovers at Af temperature. However, it is also possible to program the alloy to perform two-ways memory effect: one activated at high temperature and another at low temperature. Thus, external forces are not required to perform the shape change upon cooling the material from A to M phases (Boussu and Petitniot 2006; Hesselbach 2007; Cianchetti 2013).

The difference between the transition temperatures when heating and cooling defines the hysteresis of the process. This can be an important parameter for materials' selection in regards to small hysteresis performing faster actuation and larger

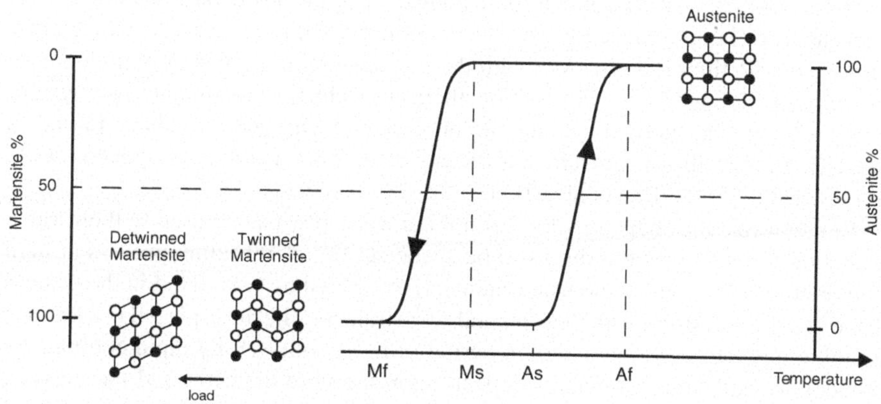

Fig. 5.3 Shape memory phase transitions

hysteresis retaining the pre-programed shape for a longer period (Honkala 2006; Liu 2010).

SMAs also present behavior called superelasticity or pseudoelasticity. This property is observed when, without any temperature change, the alloy is able to return to its original shape after a mechanical strain has been applied and is then withdrawn (Kumar and Lagoudas 2008; Cho 2010).

Moreover, SMAs also demonstrate high damping capacities in comparison with other materials. They are able to reduce mechanical vibration or to absorb shocks, as a result of the high internal frictions exhibited during phase transformation and in the M phase (Boussu and Petitniot 2006).

Since the SME was first observed in a gold-cadmium (AuCd) alloy, in 1932 by Chang and Read, several SMAs have been developed (Otsuka and Wayman 1998). Among different alloys' compositions, the SMA which has gained more attention is Nitinol, a nickel and titanium (NiTi) alloy, discovered by Buehler and co-workers in 1963, called NITINOL in honor of its discovery at the Naval Ordnance Laboratory (NOL) (Kumar and Lagoudas 2008).

This alloy presents increased properties in relation to other SMAs and research in this area has led to a great variety of applications. Available with diverse transition temperatures and forms (wire, tube, strip and sheet), Nitinol exhibits strong SME, pseudoelastic behavior, high damping capacity as well as high resistance to corrosion and abrasion. It is non-toxic, biocompatible and it is less expensive than other SMAs (Honkala 2006; Ritter 2007; Elahinia 2016). Due to the characteristics cited and taking into consideration the objective to integrate SMAs in textile substrates, Nitinol wire was the alloy selected to be addressed in this research.

The most common Nitinol proportion is near-equatomic, approximately 49% nickel by weight, but can be altered in order to attain specific properties. An increase in nickel by up to 1%, decreases the transformation temperature and increase the yield strength in the A phase. The combination of Nitinol with small quantities of other elements is also mentioned to customize specific properties, such as copper (Cu) and iron (Fe) to reduce hysteresis and nobium (Nb) to increase As temperature (Boussu and Petitniot 2006). Although the mechanical properties of Nitinol SMAs are primarily defined by its composition, thermo-mechanical treatments, ageing after solution treatment, thermal cycling and processing techniques are also reported as interfering in the alloy characteristics (Zheng et al. 2004; Hartl and Lagoudas 2008; Nurveren et al. 2008; Rao et al. 2015).

To memorize the shape that the Nitinol recovers to when exposed to their transformation temperature, it requires to be shape set by an annealing process at high temperatures. Previous to the heat treatment, the alloy has to be fixed in the desired shape in a die or fixture, with the wire ends constrained. The die is placed in a furnace at high temperature during a certain period of time, followed by rapid cooling. As temperature and time parameters depend upon the wire diameter and the mass of the die, without the existence of an accurate guidance for their definition, several authors suggest that experimentation is required (Hesselbach 2007; Liu et al. 2008; Rao et al. 2015). However, the common temperature recommended range between 400 and 550 °C and duration should take into consideration that long periods of time can interfere in the alloys' properties, Rao et al. (2015) suggest between 5 and 30 min.

The process described, refers to the one-way memory effect. The annealing treatment for two-ways memory effect is more complex and involves several thermo-mechanical cycles along a specific loading path (Hesselbach 2007; Lexcellent 2013). These repeated cycles significantly influence the alloys' microstructure and consequently their behavior: the recovered strain reduces up to 2%; there is a decrease of the transformation forces; and the memory can be lost with a slight overheating (Wan and Stylios 2007; Rottiers et al. 2011; Bengisu and Ferrara 2018).

During the heat treatment processes, Nitinol acquires an oxide layer that can influence corrosion resistance, joining ability besides other properties. Diverse surface modification processes can be applied to customize their qualities to the application requirements. The removal of these oxides can be developed by mechanical or electro-polishing, chemical etching, sandblasting, passivation and slurry ID cleaning (Shabalovskaya et al. 2008; Memry n.d.-c; NDC n.d.). Moreover, it is also possible to impart the alloy with another materials' layer, through coating and plating finishings (Memry n.d.-c).

The oxide formations can also affect the Nitinol electrical resistance. Resistive heating (or Joule effect) is a method widely used to induce the alloys' thermal activation and, when the alloys are joined together or to other materials without a proper surface modification, the oxide can cause electrical resistance issues in the connection areas (Rao et al. 2015). In addition, several authors have also researched Nitinol electrical resistance, in regards to its variation upon different phase transitions (Wu et al. 2000; Novák et al. 2008; Song et al. 2011).

Joining Nitinol components together or with other materials in a reliable way has a significant role for SMA applications. Taking into consideration each application, the reported methods include: mechanical techniques as crimping or swaging; soldering using preferentially silver bearing solders combined with aggressive flux; welding through TIG, laser, e-beam, resistance or plasma techniques; and adhesives, cyanoacrylates or epoxies (Šalej et al. 2011; Lah et al. 2016; Memry n.d.-a).

Furthermore, Nitinol alloys are commercially available within different specifications. For textile integration, Nitinol wires with round sections are the most commonly used, comprising of different thicknesses, variation temperatures, surface finishes and states such as: cold work or as drawn, straight and spring shape annealed (Memry n.d.-b). It is also possible to purchase Nitinol with a pre-memorized behavior of contraction and relaxation. This alloy, under the trade name FLEXINOL®, it is known as muscle wire (Dynalloy n.d.).

5.3 Shape Memory Textiles

SMAs material and behavior properties challenge new perspectives towards innovative and reliable applications, as well as revealing potential to supersede other materials and technologies in specific functions such as traditional hydraulic, pneumatic and motor actuators (Lawson 2016). Over the past decades, SMAs have been implemented in technical domains, namely automotive, aerospace, robotics and biomedical. More recently, they have become an exponent of research and development in textiles where, besides functional possibilities, they also explore enhanced aesthetics and interactions.

Early developments in shape memory textiles have focused on functional protective clothing. In the late nineties, an intelligent suit for firefighters was developed, incorporating Nitinol springs placed in between two fabric layers whose shape change creates an insulating air layer (Congalton 1999). The application of pseudoelastic and damping effect of SMAs has also been studied in ballistics, GEMTEX have been researching woven substrates of Nitinol combined with different materials such as high tenacity polyethylene (PE) and aramids to improve high-speed impact resistance (Boussu and Petitniot 2006).

SMAs have also found application in women's brassieres. Nitinol underwire inserted in bra pads provide increased comfort in comparison with conventional steel underwires, due the lower elastic modulus and by resisting to permanent deformation (Wu and Schetky 2000).

In 2000, Leenders explored the shape change behavior by stitching SMA wires pre-programmed with different shapes in woven textiles. The effects achieved by shrinking, creasing or rolling were applied on certain areas of shirts and jackets, creating change in size and texture (Leenders 2000).

Oricalco fabric, developed in 2001 by Corpo Nove through its spin-off firm Grado Zero Espace, was the first example of Nitinol integration in a woven textile substrate. It was used to manufacture a shirt that in response to heat, recovered its original shape, having the rolled up sleeves acquire a new size according to their pre-programmed geometry (Quinn 2010; Grado Zero Espace n.d.).

The fashion designer Hussein Chalayan, known as one of the pioneers of merging technology with fashion, presented a shape memory dress in his Before Minus Now, Spring/Summer 2000 collection. The dress with SMA stitched in its skirt, instead of being stimulated with thermal radiation, is heated with electrical current that produces the phase transformation and therefore changes the skirt volume (Lee 2005).

Design research studio XS Labs also developed shape memory garments activated electrically. In 2005, they have experimented to pre-program Nitinol in different configurations, such as curves and zigzags. However, they found that with coil or spring shape a greater performance was attained (Berzowska and Coelho 2005). On the Vilkas dress, the Nitinol was stitched on cotton and the SME caused the fabric to wrinkle, thus revealing the wearer's knee. The Kukkia dress has felt and silk flowers with Nitinol stitched to each petal to achieve an open and closed effect. Later, they developed the Skorpions collection to explore anthropomorphic qualities in garments through textile movement. The Nitinol was hand-stitched and embroidered to perform aesthetic experiences (Berzowska et al. 2007; Seymour 2008).

Coelho (2008) also relied on SMA incorporation with felt. Focusing on human-computer interactions and the shape change properties, he developed interfaces that study topological, textural and permeable transformations and researched how to use form transformation as a tool for communication and expression (Coelho and Zigelbaum 2011).

In 2008/2009, Schäth developed Emotion: Outsourcing, a shape-shifting garment that addresses the wearer's senses and feelings via SMA and integrated sensors. By the jacket hood changes, the author intended to make the wearer aware of mood variation (Seymour 2010).

Regarding yarn spinning with SMA and following integration on textile structures, (Chan et al. 2002) conducted research focused on aesthetics and design. They have produced a variety of yarns with pre-programmed SMA as a core element, wrapped with conventional fibers. They found that the twist level and rollers' speed were important parameters to be optimized in order to avoid SMA protruding from the yarn and reducing SME constraints. Initial experiments using the yarns for knitting, exposed limitations due to the lack of extensibility of the SMA yarn and propensity to deform when knitted. Better results were achieved introducing SMA yarn formations in selected areas of the knitted structure (Winchester and Stylios 2003). Tests aiming at weaving, with SMA yarns in the weft, revealed yarn end breaking when tension was too tight or slack. Samples were developed with SMA yarn formation with coil shapes, floating on the surface. A range of textures and sculptured woven and knitted fabrics were designed (Stylios 2006; Vili 2007).

Wan and Stylios (2007) have conducted experiments on the annealing process of Nitinol wire, for one-way and two-ways memory effect and their integration in woven structures. They have verified that the effect of repeated treatment cycles for two-ways shape change have significantly decreased the amount of the alloys' recoverable strain. Additionally, they suggest the implementation of a bias mechanism in the textiles, to overcome this limitation.

Concerning experiments with SMA placed between two layers of fabric, Laschuk and Souto (2008) have developed SMA bras, which in near contact with skin recover their memorized shape. They have produced bra pads with knitted Nitinol and cotton yarn. Conventional yarn was applied in order to provide a better processability. After heat treatment, using a ceramic die, the cotton materials burn and are removed from the samples, resulting in 100% SMA structures. They also developed woven pads; however, in this case, it was not necessary to include textile fibers into the process. The woven samples achieved better performance (Laschuk 2008).

Dyer (2010) alludes to the integration of Nitinol in woven fabrics. His research focused on the influence of textile yarn, woven structure, cloth set and Nitinol selection in the mechanical shape transfer and interfacial relationships. Results have demonstrated that lighter and tighter structures attain greater shape transfer than heavier or looser samples and plain weave samples with high density of threads required a higher loading to initiate the removal of Nitinol from the textile structure. The slippage phenomena was also studied by Vasile et al. (2010) and results demonstrated lower friction levels in samples with lower number of picks/cm and relatively high yarn diameter.

A SMA Velcro system was developed in 2011, presenting Nitinol hook fasteners on the male strip. Through electrical activation the hooks open and close when they cool down, providing a silent Velcro system (Vokoun et al. 2011).

Techno Naturology, a hybrid word that combines nature and technology, was the concept that Elaine Ling addressed in her investigation. Focusing on hybrid materialization, through diverse techniques such as weaving, laser-cutting and laminated surfaces, she studied and applied natural and man-made responsive materials, namely wood, SMAs and SMPs. Ling's collections achieved mimicked tectonic movements in response to changes in temperature and moisture, reflecting the micro behavior of how nature becomes evident (Dieffenbacher 2013).

Michalak and Krucińska (2015) have researched the two-ways memory effect to develop a controllable heat insulation textile. The Nitinol was annealed in spring spirals shape that changed their length when activated and when cooled down. The process was patented (Michalak et al. 2013). The Nitinol actuators were placed in between a conductive nonwoven inner layer and 2 non-conductive outer layers, attaining an interlayer width increase of approximately 2.5 mm.

Lah et al. (2016) have researched the two-ways memory effect in a knitted structure. After producing a 100% Nitinol (0.2 mm) hand weft knitted, the annealing process began with the prototype mounted in a half-sphere form, placed in a furnace at 500 °C for 30 min and air-cooled to a temperature below 22 °C for 20 min. The prototype was then mounted in a flat form and annealed at 75 °C for 10 min and again cooled at below 22 °C for 20 min. The authors mentioned that this process was repeated more than 10 times, attaining the two-ways shape memory. As in Michalak et al., this research does not allude for the Nitinol recovered strains and transformation forces within the two-ways activation cycles.

The projects discussed address diverse approaches regarding SME training, integration of SMAs in textiles substrates and activation method. The materials' systems studied are diverse and comprise different concepts, processes and product applications, attaining milestones in shape memory textiles development.

The researchers also share common issues. The Nitinol shape set undergoes a thermo-mechanical process that can be optimized for one-way or two-ways shape memory. However, the repeated cycles necessary to impart the alloy to memorize two shapes, is reported to decrease the alloys' shape change properties and a bias mechanism is suggested as an alternative for the development of two shapes actuation (Hesselbach 2007; Wan and Stylios 2007; Rottiers et al. 2011; Cianchetti 2013; Elahinia 2016). Furthermore, SMAs memorize their shape at considerably higher temperatures than those supported by textile fibers. In this sense, they must be processed prior to their integration in fabric manufacturing, if a 100% SMA textile is not the goal (Stylios 2006; Laschuk and Souto 2008; Ugur 2013).

Another aspect is related to the fact that textiles require a certain amount of stretch in the yarns to be processed. The low strain properties of SMAs represent a challenge when the material is to be spun, woven or knitted. Textile structures' characteristics and materials interaction can also constrain the SME, reducing the ability to perform the shape change (Chan et al. 2002; Stylios 2006; Vili 2007; Jani et al. 2014). In woven structures, a greater number of interlacings between the Nitinol and perpendicular warp threads has improved control and integration of Nitinol in the textile (Dyer 2010). The ability of the SMA wire to be pulled out from the host structure was also highlighted and, besides greater resistance was found in samples with higher weft density, they still show tendency to easily slide (Dyer 2010; Vasile et al. 2010).

Concerning the activation methods, different approaches have been taken. As a thermo-responsive material, Nitinol alloys can assume their pre-programmed shape by a direct response to room temperature or thermal radiation from nearby bodies (Leenders 2000; Grado Zero Espace n.d.; Laschuk and Souto 2008). As a conductive material, it is also possible to activate the Nitinol alloys through resistive heating. By this process, the shape change can be programmed and can have integrated sensors in order to explore SME in response to diverse stimuli that activate SMA heating (Lee 2005; Berzowska et al. 2007; Seymour 2010; Vokoun et al. 2011). The SME analyzed within the projects fulfils changes in textile texture, volume, contractions and creasing, rolling up and forward and backward movement.

References

Adiguzel O (2015) The role of twinned and detwinned structures on memory behaviour of shape memory alloys. Adv Mater Res 1105:78–82

Bengisu M, Ferrara M (2018) Materials that move: smart materials, intelligent design. Springer International Publishing, Cham

Berzowska J, Coelho M (2005) Kukkia and Vilkas: kinetic electronic garments. In: 9th IEEE international symposium on wearable computers, 18–21 October 2005, Osaka, Japan

Berzowska J, Mainstone D, Bromley M, Coelho M, Gauthier D, Raymond F, Boxer V (2007) Skorpions: kinetic electronic garments. In: 9th international conference on ubiquitous computing, Innsbruck, Austria

Boussu F, Petitniot J (2006) Development of shape memory alloy fabrics for composite structures. In: Mattila HR (ed) Intelligent textiles and clothing. Woodhead, Cambridge

Chan YYF, Winchester RCC, Wan TR, Stylios GK (2002) The concept of aesthetic intelligence of textile fabrics and their application for interior and apparel. In: IFFTI international conference, 7–9 November 2002, Hong Kong, pp 458–471

Cho CG (2010) shape memory material. In: Cho G (ed) Smart clothing: technology and applications. CRC Press, Boca Raton, FL

Cianchetti M (2013) Fundamentals on the use of shape memory alloys in soft robotics. In: Interdisciplinary mechatronics. Wiley

Coelho M (2008) Materials of interaction: responsive materials in the design of transformable interactive surfaces. Master thesis, Massachusetts Institute of Technology

Coelho M, Zigelbaum J (2011) Shape-changing interfaces. Pers Ubiquitous Comput 15(2):161–173. https://doi.org/10.1007/s00779-010-0311-y

Congalton D (1999) Shape memory alloys for use in thermally activated clothing, protection against flame and heat. Fire Mater 23:223–226

Dieffenbacher F (2013) Fashion thinking: creative approaches to the design process. AVA Academia, Lausanne

Dileonardo-Parker B (2016) Six simple twists: the pleat pattern approach to origami tessellation design. CRC Press, Boca Raton

Dyer PE (2010) Dynamic control of active textiles: the integration of nickel-titanium shape memory alloys and the manipulation of woven structures. PhD thesis, University of Brighton

Dynalloy (n.d.) Introduction To FLEXINOL® Actuator Wire [Online]. http://www.dynalloy.com/flexinol.php. Accessed 20 March 2013

Edwards A, Yan H (2014) DNA origami. In: Kjems J, Ferapontova E, Gothelf KV (eds) Nucleic acid nanotechnology. Springer, Berlin
Elahinia MH (2016) Shape memory alloy actuators: design, fabrication, and experimental evaluation. Wiley, Chichester, West Sussex
Ge Q, Dunn CK, Qi HJ, Dunn ML (2014) Active origami by 4D printing. Smart Mater Struct 23:094007. https://doi.org/10.1088/0964-1726/23/9/094007
Gjerde E (2009) Origami tessellations: awe-inspiring geometric designs. A K Peters, Massachusetts
Gök MO, Bilir MZ, Gürcüm BH (2015) Shape-memory applications in textile design. Procedia Soc Behav Sci 195:2160–2169. https://doi.org/10.1016/j.sbspro.2015.06.283
Goldstein BE (2010) Sensation and perception. Wadsworth Cengage Learning, Belmont, CA
Gordon IE (2004) Theories of visual perception. Psychology Press, Taylor and Francis Group, Hove and New York
Grado Zero Espace (n.d.) Oricalco [Online]. http://www.gradozero.eu/gzenew/index.php?pg=oricalco&lang=en. Accessed 2 Sept
Hartl DJ, Lagoudas DC (2008) Thermomechanical characterization of shape memory alloy materials. In: Lagoudas DC (ed) Shape memory alloys: modeling and engineering applications. Springer, New York
Hartl DJ, Lagoudas DC, Malak R, Frecker M, Ounaies Z (2014) Active materials and structures for origami engineering. Smart Mater Struct 23:090201 (2 pp). https://doi.org/10.1088/0964-1726/23/9/090201
Hartl DJ, Lane K, Malak R (2012) Computational design of a reconfigurable origami space structure incorporating shape memory alloy thin films. In: Conference on smart materials adaptive structures, and intelligent systems, 2012, Georgia, USA
Hawkes E, An B, Benbernou NM, Tanaka H, Kim S, Demaine ED, Rus D, Wood RJ (2010) Programmable matter by folding. Proc Natl Acad Sci 107:12441–12445. https://doi.org/10.1073/pnas.0914069107
Hensel M, Menges A (2006) Morpho-ecologies. Architectural Association, London
Hensel M, Menges A, Weinstock M (2006) Techniques and technologies in morphogenetic design. Wiley-Academy, London
Hesselbach J (2007) Shape memory actuators. In: Janocha H (ed) Adaptronics and smart structures: basics, materials, design and applications. Springer, Berlin
Honkala M (2006) Introduction to shape memory materials. In: Mattila HR (ed) Intelligent textiles and clothing. Woodhead, Cambridge
Hu J (2007) Shape memory polymers and textiles. Woodhead in Association with Textile Institute, Cambridge
Hu J, Meng Q (2011) Functional shape memory textiles. In: Pan N, Sun G (eds) Functional textiles for improved performance, protection and health. Woodhead Publishing, Oxford
Huang WM, Ding Z, Wang CC, Wei J, Zhao Y, Purnawali H (2010) Shape memory materials. Mater Today 13(7–8):54–61. https://doi.org/10.1016/S1369-7021(10)70128-0
Huang WM, Yang B, An L, Li C, Chan YS (2005) Water-driven programmable polyurethane shape memory polymer: demonstration and mechanism. Appl Phys Lett 86:114105. https://doi.org/10.1063/1.1880448
Humbeeck JV (2009) Shape memory Alloys. In: Schwartz MM (ed) Smart materials. CRC Press, Boca Raton
Jackson P (2011) Folding techniques for designers: from sheet to form. Laurence King Publishing, London
Jani JM, Leary M, Subic A, Gibson MA (2014) A review of shape memory alloy research, applications and opportunities. Mater Des 56:1078–1113. https://doi.org/10.1016/j.matdes.2013.11.084
Koizumi N, Yasu K, Liu A, Sugimoto M, Inami M (2010) Animated paper: a toolkit for building moving toys. Comput Entertain 8:1–16. https://doi.org/10.1145/1899687.1899689
Kumar PK, Lagoudas DC (2008) Introduction to shape memory alloys. In: Lagoudas DC (ed) Shape memory alloys: modeling and engineering applications. Springer, New York

Lah AŠ, Fajfar P, Lavrič Z, Bukošek V, Rijavec T (2016) Preparation of shape memory NiTiNOL filaments for smart textiles. Tekstilec 59(2):168–174. https://doi.org/10.14502/Tekstilec2016.59.168-174

Lang RJ (2012) Origami design secrets: mathematical methods for an ancient art. CRC Press, Boca Raton

Lang RJ (2018) Twists, tilings and tessellations: mathematical methods for geometric origami. CRC Press, Boca Raton

Laschuk T (2008) Application of smart textiles to fashion textile products. Master thesis, University of Minho

Laschuk T, Souto AP (2008) Incorporation of SMA technologies in fashion underwear apparel. Ambience 08, 2–3 June 2008, pp 215–218

Lawson BC (2016) Shape memory alloys, muscle wires and robotics. S.I., CreateSpace

Lee S (2005) Fashioning the future: tomorrow's wardrobe. Thames & Hudson, London

Leenders L (2000) Shape memory textiles [Online]. http://photorepeats.com/en/work/shape-memory-textiles/. Accessed 3 Sept

Lendlein A, Jiang H, Junger O, Langer R (2005) Light-induced shape-memory polymers. Nature 434:879–882. https://doi.org/10.1038/nature03496

Lexcellent C (2013) Shape-memory alloys handbook. ISTE Ltd/John Wiley and Sons Inc, Hoboken, NJ

Li S, Fang H, Sadeghi S, Bhovad P, Wang K-W (2019) Architected origami materials: how folding creates sophisticated mechanical properties. Adv Mater 31:1805282. https://doi.org/10.1002/adma.201805282

Liu X, Wang Y, Yang D, Qi M (2008) The effect of ageing treatment on shape-setting and superelasticity of a nitinol stent. Mater Charact 59:402–406. https://doi.org/10.1016/j.matchar.2007.02.007

Liu Y (2010) Some factors affecting the transformation hysteresis in shape memory alloys. In: Chen HR (ed) Shape memory alloys: manufacture, properties and applications. Nova Science Publishers, New York

May M (2007) Sensation and perception. Chelsea House Publishers, New York

Memry (n.d.-a) Joining of Nitinol [Online]. http://www.memry.com/nitinol-iq/nitinol-fundamentals/joining. Accessed 11 Jan 2014

Memry (n.d.-b) Nitinol wire [Online]. http://www.memry.com/products-services/material/wire. Accessed 20 March 2013

Memry (n.d.-c) Surface finishing, coating & plating of nitinol [Online]. http://www.memry.com/nitinol-iq/nitinol-fundamentals/surface-finishing-coating-plating. Accessed 11 Jan 2014

Michalak M, Bilska J, Krucińska I, Zięba J, Gniotek K (2013) Siłownik termiczny (thermal actuator)

Michalak M, Krucińska I (2015) A smart fabric with increased insulating properties. Text Res J. https://doi.org/10.1177/0040517515581585

Mitani J (2011) A method for designing crease patterns for flat-foldable origami with numerical optimization. J Geom Graph 15(2):195–201

Miura K (1994) Map Fold a La Miura Style, its physical characteristics and application to the space science. In: Takati R (ed) Research of pattern formation. KTK Scientific Publisher, Tokyo

Mu J, Hou C, Wang H, Li Y, Zhang Q, Zhu M (2015) Origami-inspired active graphene-based paper for programmable instant self-folding walking devices. Sci Adv 1:e1500533. https://doi.org/10.1126/sciadv.1500533%JScienceAdvances

Ndc (n.d.) Nitinol facts: surface finishes [Online]. www.nitinol.com/nitinol-university/nitinol-facts. Accessed 11 Jan 2014

Nefs HT (2008) On the visual appearance of objects. In: Schifferstein H, Hekkert P (eds) Product experience. Elsevier, San Diego, CA

Novák V, Šittner P, Dayananda GN, Braz-Fernandes FM, Maheshc KK (2008) Electric resistance variation of NiTi shape memory alloy wires in thermomechanical tests: experiments and simulation. Mater Sci Eng 481–482:127–133. https://doi.org/10.1016/j.msea.2007.02.162

Nurveren K, Akdoğan A, Huang WM (2008) Evolution of transformation characteristics with heating/cooling rate in NiTi shape memory alloys. J Mater Process Technol 196:129–134. https://doi.org/10.1016/j.jmatprotec.2007.05.015
Onal CD, Tolley MT, Wood RJ, Rus D (2015) Origami-inspired printed robots. IEEE/ASME Trans Mechatron 20:2214–2221. https://doi.org/10.1109/tmech.2014.2369854
Otsuka K, Kakeshita T (2002) Science and technology of shape-memory alloys: new developments. MRS Bull 27:01–100. https://doi.org/10.1557/mrs2002.43
Otsuka K, Wayman CM (1998) Shape memory materials. Cambridge University Press, Cambridge
Paik JK, An B, Rus D, Wood RJ (2011) Robotic origamis: self-morphing modular robots. In: II international conference on morphological computation, 2011 Venice, Italy
Peraza Hernandez E, Hartl DJ, Lagoudas DC (2019) Active origami: modeling, design and applications. Cham, Springer
Peraza-Hernandez EA, Hartl DJ, Malak Jr RJ, Lagoudas DC (2014) Origami-inspired active structures: a synthesis and review. Smart Mater Struct 23:094001 (28 pp)
Qi J, Buechley L (2012) Animating paper using shape memory alloys. In: Proceedings of the SIGCHI conference on human factors in computing systems, 2012 Austin, Texas, USA. ACM, pp 749–752
Quinn B (2010) Textile futures: fashion, design and technology. Berg, Oxford
Rao A, Srinivasa AR, Reddy JN (2015) Design of shape memory alloy (SMA) actuators. Springer International Publishing, Cham
Rasmussen MK, Pedersen EW, Petersen MG, Hornb K (2012) Shape-changing interfaces: a review of the design space and open research questions. In: Proceedings of the SIGCHI conference on human factors in computing systems, Austin, Texas, USA. ACM
Richards A (2012) Weaving textiles that shape themselves. Crowood, Ramsbury
Ritter A (2007) Smart materials in architecture, interior architecture and design. Basel, Birkhäuser
Rottiers W, Van Den Broeck L, Peeters C, Arras P (2011) Shape memory materials and their applications. Korolev's Readings, 2011 Samara, Russia. Samara State Aerospace University
Rutzky J, Palmer CK (2011) Shadowfolds: surprisingly easy-to-make geometric designs in fabric. Kodansha International, New York
Ryu J, D'amato M, Cui X, Long KN, Qi HJ, Dunn ML (2012) Photo-origami—bending and folding polymers with light. Appl Phys Lett 100:161908. https://doi.org/10.1063/1.3700719
Šalej A, Bombač D, Fajfar P, Rijavec T (2011) Application of Nickel-Titanium alloys (NiTi-NOL) in textiles and clothes. Tekst J TextE Cloth Technol 60:123–131
Saul G, Xu C, Gross MD (2010) End-user design and fabrication of interactive paper devices. In: TEI'10 fourth international conference on tangible, embedded, and embodied interaction, 2010 Cambridge, Massachusetts, USA, pp 205–212
Schmelzeisen D, Koch H, Pastore C, Gries T (2018) 4D textiles: hybrid textile structures that can change structural form with time by 3D printing. In: Kyosev Y, Mahltig B, Schwarz-Pfeiffer A (eds) Narrow and smart textiles. Springer, Cham
Seymour S (2008) Fashionable technology: the intersection of design, fashion, science, and technology. Springer, Wien
Seymour S (2010) Functional aesthetics: visions in fashionable technology. Springer, Wien
Shabalovskaya S, Anderegg J, Van Humbeeck J (2008) Critical overview of Nitinol surfaces and their modifications for medical applications. Acta Biomater 4:447–467. https://doi.org/10.1016/j.actbio.2008.01.013
Shimizu S (2016) Creativity is born: Issey Miyake and Reality Lab, s.l., PIE International
Sibley TQ (2015) Thinking geometrically: a survey of geometries. Mathematical Association of America, Washington, D.C.
Singer R (2013) Fabric manipulation: 150 creative sewing techniques. David & Charles Publishers, Newton Abbot
Song H, Kubica E, Gorbet R (2011) Resistance modelling of SMA wire actuators. In: Smart materials, structures & NDT in aerospace, 2011 Quebec, Canada

Stylios GK (2006) Engineering textile and clothing aesthetics using shape changing materials. In: Mattila HR (ed) Intelligent textiles and clothing. Woodhead, Cambridge
Sun L, Huang WM, Ding Z, Zhao Y, Wang CC, Purnawali H, Tang C (2012) Stimulus-responsive shape memory materials: a review. Mater Des 33:577–640. https://doi.org/10.1016/j.matdes.2011.04.065
Tachi T (2010) Origamizing polyhedral surfaces. IEEE Trans Vis Comput Graph 16:298–311. https://doi.org/10.1109/TVCG.2009.67
Tao X (2001) Smart technology for textiles and clothing—introduction and overview. In: Tao X (ed) Smart fibres, fabrics and clothing: fundamentals and applications. Woodhead, Cambridge
Ugur S (2013) Wearing embodied emotions: a practice based design research on wearable technology. Springer Publishing Company, Incorporated
Vasile S, Grabowska KE, Ciesielska IL, Githaiga J (2010) Analysis of hybrid woven fabrics with shape memory alloys wires embedded. Fibres TextEs East Eur 18(1):64–69
Vaughan LC (1997) Understanding movement. CHI'97: 548–549
Vili YYFC (2007) Investigating smart textiles based on shape memory materials. Text Res J 77:290–300. https://doi.org/10.1177/0040517507078794
Vokoun D, Sedlaák P, Frost M, Pilch J, Majtás D, Sittner P (2011) Velcro-like fasteners based on NiTi microhook-arrays. Smart Mater Struct 20. https://doi.org/10.1088/0964-1726/20/8/085027
Wada YI (2012) Memory on cloth: Shibori now. Kodansha International, New York
Wan T, Stylios GK (2007) Shape memory training for smart fabrics. Trans Inst Meas Control 29:321–336. https://doi.org/10.1177/0142331207069479
Winchester RCC, Stylios GK (2003) Designing knitted apparel by engineering the attributes of shape memory alloy. Int J Cloth Sci Technol 15:359–366. https://doi.org/10.1108/09556220310492624
Wu M, Schetky L (2000) Industrial applications for shape memory alloys. In: International conference on shape memory and superelastic technologies, 2000 Pacific Grove, California, pp 171–182
Wu XD, Fan YZ, Wu JS (2000) A study on the variations of the electrical resistance for NiTi shape memory alloy wires during the thermo-mechanical loading. Mater Des 21:511–515. https://doi.org/10.1016/S0261-3069(00)00022-4
Yeduru S (2013) Development of microactuators based on the magnetic shape memory effect. KIT Scientific Publishing, Karlsruhe, Baden
Young R, Pezzutti D, Pill S, Sharp R (2005) The language of motion in industrial design. Des Semant Form Mov: 6–12
Zhang Z, Demir KG, Gu GX (2019) Developments in 4D-printing: a review on current smart materials, technologies, and applications. Int J Smart Nano Mater. https://doi.org/10.1080/19475411.2019.1591541
Zheng Y, Cui L, Schrooten J (2004) Temperature memory effect of a nickel–titanium shape memory alloy. Appl Phys Lett 84:31–33. https://doi.org/10.1063/1.1637958

Chapter 6
Experimental Work: Shape Memory Alloys

The aim of this research section is to develop shape memory woven textiles in which dynamic behavior achieves predefined geometric morphologies. The design concept inherent to the materials research hinges on the interaction of the textile SME and light.

To perform shape change, Nitinol alloys require an annealing process to memorize the geometry they assume when exposed to their activation temperature. The heat treatment required to shape set the alloys has to be conducted at temperatures significantly higher than those supported by textile fibers. In this sense, they must be processed prior to their integration in the textile substrate.

The experimental work conducted with SMAs is covered in the following sections, comprising of preliminary analysis of light transmittance variation through the change of textile layer number, research on Nitinol annealing processes, their integration in woven substrates to develop textiles with one and two shapes behavior, electrical activation of shape change and finishing processes.

6.1 Introduction

Towards the development of shape memory woven textiles in which dynamic behavior achieves predefined geometric morphologies, a study on textile form was conducted through origami techniques. Combining foldability properties with shape change, it is possible to design textile structures with a variable number of layers, interfering in light transmittance. Initially, it was conducted a preliminary analysis of luminosity differences according to textile layer variation.

The experimental work concerned with the shape memory textile development and performance analysis has involved several studies. First, the objective was to analyze and optimize the accuracy of the shape's angles memorized in the Nitinol alloys, in relation to the predefined geometry. Having achieved an understanding of

I. D. Cabral et al., *Dynamic Light Filters*, Springer Series in Design and Innovation 3, https://doi.org/10.1007/978-3-030-39529-2_6

the associated parameters, the question raised was how to integrate SMA wires in textile substrates that can perform predefined morphologies. The experimental setup has involved the development of physical models comprised of (a) paper samples to define morphology, shape change and Nitinol wires' geometry according to the crease pattern; (b) textile samples to research manufacturing processes, morphology coherency as well as shape change behavior and characteristics.

Study samples of origami base folds and an intricate origami pattern were developed, in order to research a workflow setup for the development of shape memory textiles that change from one temporary to a memorized morphology. Textile activation through resistive heating was also studied and the results have highlighted the requirement to increase adhesion of the Nitinol within the host structure as well as electrical insulation of the alloys. Several finishing processes were studied along to the influence of a selected one in the shape change behavior.

After, the experimental work focused on the development of shape memory textiles able to perform two shape behavior through a bias mechanism. A different Nitinol integration approach was proposed in order to improve the textile behavior and the effect of changes in the Nitinol geometries was studied across samples dimensions and positioning during shape change activation.

The experiments conducted have attained an understanding of the manufacturing processes and performative variables to develop and assess the shape change textiles behavior within predefined geometric shapes.

Materials, equipment and methods applied in this chapter experimental work are described in Fig. 6.1.

6.2 Textile Morphology and Light Transmittance

For a preliminary analysis of the transmittance variation that the dynamic textile morphology may produce and for a proof-of-concept in an initial stage of the work, two experiments were conducted. First, the relation was studied using simple pile-up of textile layers. After, two origami-based textiles were developed to analyze the phenomena upon layer variation in specific areas, attained by the folding and unfolding behavior.

To understand how the amount of layers can transform the fraction of incident light that goes through them, a photographic record and a light meter measurement were conducted using piled layers of the textile substrate (Fig. 6.1).

The tests were conducted in a dark room and different amounts of textile layers were placed on the open face of a light box. For the photo record, the light box was set with the open face perpendicular to a wall and a camera set on a tripod. In the light intensity measurement, the open face of the light box was set in vertical position, 1 m from the luxmeter. The lamp was turned on and the luxmeter values recorded on three points.

For the study of layer variation upon the folding and unfolding, two textile origamis were developed. The origami models comprise of intricate patterns based

MATERIALS

Textile substrate:	50% cotton (CO) 50% polyester (PES) plain weave, 115g/m2 15,0 warp and 15,5 weft (tex), density of 41 ends/cm and 31 picks/cm
Nitinol alloys:	Saes Getters Nitinol wire in as drawn state, with three diameters: 0,2; 0,3 and 0,5 mm. Transformation temperature As +30ºC and Af +45ºC.
Finishing agents:	1 – A paste that combines Aktilem 14000 and Aktilem 13002 with a vinyl acrylic binder. Concentrations applied were 15%; 15%; 70%, respectively and viscosity measured in the viscometer was 3700cP; 2 – A paste that combines Tanacoat OMP and Acraconc 2C-N. Concentrations applied: 99,9%; 0,1%, respectively and viscosity measured of 2300cP; 3 – PTFE; 4 - Baygard AFF 300% 01; 5 – Insulating agent with an alkyd resin base.
TC pigments:	TC pigments Varioterm AC base 27ºC - orange, red and black
Binder:	Gilaba vinyl acrylic binder

EQUIPMENT & METHODS

Nitinol annealing processes were developed in a Naber D-2804 furnace.

The weaving processes were developed in three looms:

a) Initially a table handloom was used, set up with 3/24 CO yarn (148,7 Tex) warp, 16 ends/cm. In the weft, the same yarn was used and samples presented 12 picks/cm. The width of the warp was 12 cm;

b) After initial experiments, the samples were woven in an electronic Jacquard loom with 3456 hooks and a Bonas Series 200 Controller. During this research time frame, the loom warp was changed. The first setup had a 30,8 Tex CO (unbleached) yarn with a density of 38 ends/cm in the warp and in the weft it was used 17,7 Tex PES yarn accomplishing 11 picks/cm. The second setup presented a 17 Tex PES texturized yarn with 3 x 30 cm of black, white and grey yarn colours width and a density of 38 ends/cm. A 25,0 Tex CO (bleached) yarn was used in the weft, with a density of 11 picks/cm. In the experimental section, the respective loom setups will be reported.

In this loom, it was not possible to program the interruption of the weft insertion according to the predefined geometry of the Nitinol integration. The weaving process was conducted in slow mode, in order to ensure the number of textile picks in between the SMAs wires;

c) Samples woven through a programmed interruption of the weft insertion were developed in the weaving laboratory at The Swedish School of Textiles, University of Borås, Sweden. The equipment used was a Jacquard Vamatex SD 1701 loom of 5280 hooks, with a Grosse controller and ScotWeave CAD software. The warp had 41,2 Tex CO (unbleached) yarn with a density of 33 ends/cm and the weft had 14 picks/cm with 14,7 Tex PES yarn. The warp width was 160 cm and the pattern definition contemplates a maximum of 1320 ends within 4 times repetition and a maximum of 2000 picks. The warp yarn was also substituted during this research path, for a 38,7 Tex bleached CO yarn. The remaining weaving characteristics were equal.

Fig. 6.1 Materials, equipment and methods applied in the experimental work

Finishing agents were applied through screen printing, knife coating, impregnation and pulverization (spraying). The pastes viscosity was measured in a Brookfield DV-E Viscometer with a LV spindle S4. Screen printing was conducted in a Zimmer Mini MDF R541 table, with following settings: magnetic field level 3, application speed 50% of the graded dial and higher rod diameter parameter. A 12 mm metallic rod-squeegee and an open screen with 101,6 TPI were used. Knife coating processes were conducted using a Mathis laboratory coating device type SV with a knife-over-roll cylinder. Each sample was placed in the pin frame, the micrometre was set at 0,5 mm and each paste was applied with a coating knife shape A.

After drying, samples were thermo set in a laboratory oven with time and duration parameters according to the finishing agents applied: 1 and 2 - 150ºC during 5 minutes; 4 - 160ºC during 1 minute; 3 and 5 - did not required to be thermo set. Samples screen printed with TC pigments used equal printing parameters and were thermo set at 150ºC during 3 minutes

To analyse the light transmittance variation, a light box with a 28 x 18 cm aperture was used to place the textile samples and the light intensity was measured with a DeltaOhm – HD 2302.0 luxmeter.

Shape change activation through resistive heating was conducted with DC power supply. Two devices have been used: TENMA 72-8695 power supply and Velleman LABPS3003SM power supply.

To study samples' temperature during resistive heating activation, thermal images were recorded with a Testo 876 Infrared camera and analysed with Testo IRSoft software.

A Hounsfield H10KS dynamometer (strengh tester) was used to study the resistance of the Nitinol wire to be removed from the textile host structure. The evaluation was conducted through a Breaking Strength and Elongation test at a loading rate of 5cm/min.

The measurement of the samples thickness was conducted with a Mitutoyo 7321 Thickness Gauge.

Fig. 6.1 (continued)

on the waterbomb and squaretwist crease patterns (Fig. 6.2). They are origami base folds and were chosen for this experiment and for the study of the SMA integration in textiles, due the different level of complexity that they might present for shape memory textile development and performance. Waterbomb is based on pleats while squaretwist displays pleats and twist. The origami patterns that can be designed with these base folds are limitless.

The unfolding of the samples was carried out manually and focused on the creation of spaces between the folds, instead of achieving the complete unfolding of the textile surface. To evaluate light transmittance, the luxmeter was placed 5 cm above the light box top, pointing to a white surface placed 1.0 m in front of the light box, in a first measurement cycle and 0.5 m in the second. Light intensity values were recorded on three points.

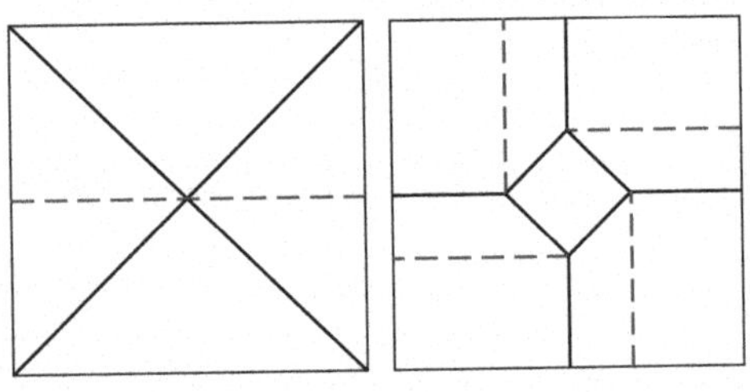

Fig. 6.2 Waterbomb and Squaretwist crease pattern (left and right, respectively) (Cabral et al. 2015)

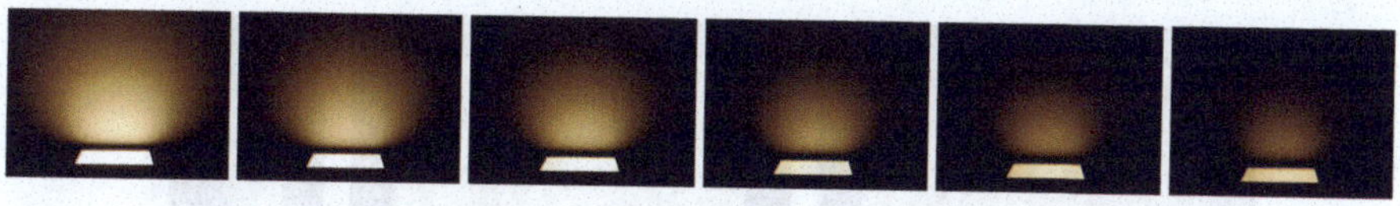

Fig. 6.3 Pile-up of textiles from one to six layers (left to right, respectively)

The images of the pile-up layers from one to six are presented in Fig. 6.3. Light projection on the wall allows visualization of the relation between the blur size of the projected light and light intensities. The decrease of light transmittance among layer number is perceived and, comparing one textile layer (image on the left) with six pile-up layers (image on the right), the light projection areas are distinct.

Table 6.1 shows the mean values of measured light intensities attained with the piled layers. According to the results, it is possible to conclude that the variation of the number of textile layers achieves a considerable change of light transmittance. From one to two layers, the reduction was about 43%, additional layers produced a decrease of percentual luminosity difference. The values attained until five layers were considered more expressive to design dynamic light filters and presented a base framework for the following experimental sections with origami samples.

The resulting origami model presented in Fig. 6.4 consists of pleated squares created by waterbomb folds in each square vertex. The unfolding was carried out by pulling up the center square, causing a volume increase with a partial unfolding of the waterbomb folds. In Fig. 6.5, the model is composed of squaretwist folds, regularly aligned. This origami presents vertical and horizontal tabs created by the

Table 6.1 Mean values of light intensities

No. of layers	Light intensity (lx)	Difference (%) 1 layer to × layers
1	13.38	–
2	7.62	43
3	5.17	61
4	3.83	71
5	2.64	80
6	2.09	84

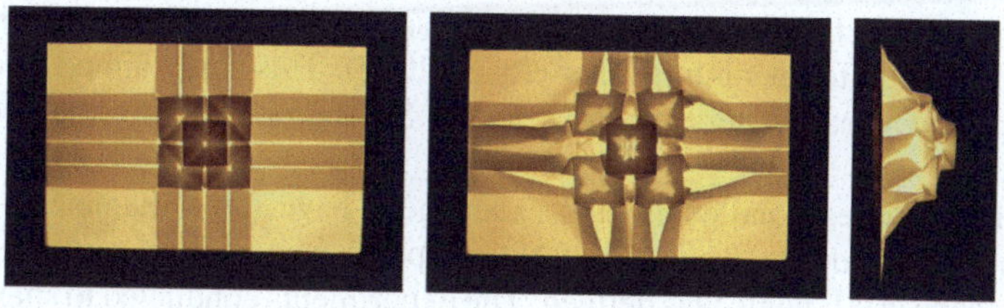

Fig. 6.4 Study model waterbomb fold based: folded (left) and unfolded (center and right) (Cabral et al. 2015)

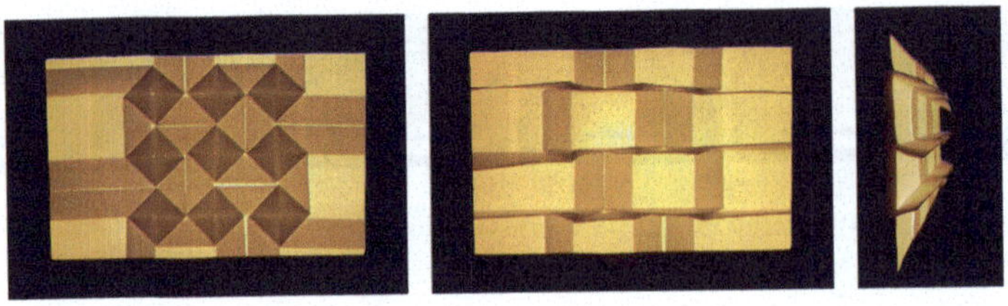

Fig. 6.5 Study model squaretwist fold based: folded (left) and unfolded (center and right) (Cabral et al. 2015)

Table 6.2 Mean values of light intensities

Sample	Distance (m)	Light intensity (lx)		Luminosity difference (%)
		Folded	Unfolded	
Waterbomb based	1.00	2.54	6.55	61.00
	0.50	5.05	11.73	57.00
Squaretwist based	1.00	1.42	5.38	74.00
	0.50	2.18	7.44	71.00

twist of the squares. The shape change was produced by folding the horizontal tabs orthogonally to the origami surface.

Table 6.2 presents the measured light intensities of the folded and unfolded models. The results show a difference of light transmittance between 61 and 57% with the waterbomb based sample and between 74 and 71% in the squaretwist based sample. By direct observation, it was also perceived the change of the ambient light intensity.

The values of the experiments conducted relate to the specific characteristics of the textile substrate applied and the origami crease patterns. However, they set an initial understanding of the transmittance variation that might be achieved with different number of textile layers, presenting the possibility to create dynamic light with textiles structures whose morphology displays shape change.

6.3 Nitinol Shape Set

To train the SME of the Nitinol wire, it is necessary to fix it in the desired shape and condition it to high temperatures (Hesselbach 2007). This study aimed to optimize the heat treatment parameters for the selected SMAs, considering the geometries defined.

The Nitinol wires handled in this experiment have three diameters: 0.2; 0.3 and 0.5 mm. To study the accuracy of Nitinol performance in reacquiring the programmed shape, a 90° angle was defined. The experiments conducted to shape set the Nitinol, took into consideration the annealing parameters indicated by the supplier—temperature and duration of the heat treatment. It was advised to apply a temperature

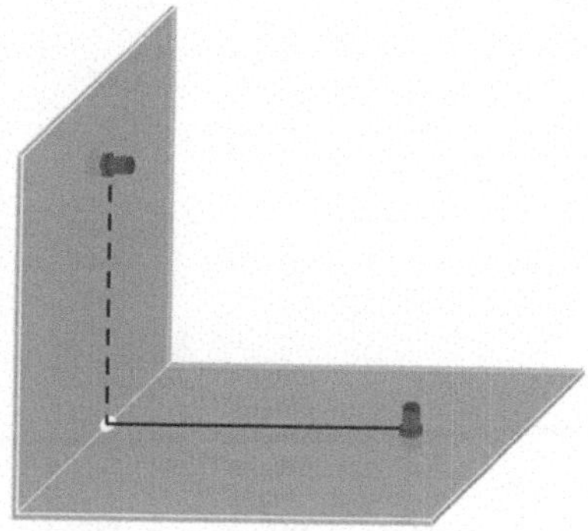

Fig. 6.6 Nitinol die setup (Cabral et al. 2015)

between 450 and 550 °C. Duration was suggested to be tested, highlighting that long periods of time could interfere with the wire's fatigue and the ability to perform repeatable accurate predefined shape changes throughout several cycles.

A stainless steel sheet with 1 mm thickness was used to produce a die in order to create the desired shape. The exact location of the wire crease was set with an orifice through which the wire is led and the wire ends were fixed with screws and nuts (Fig. 6.6). The distance between the screws and the crease orifice is 5.5 cm and after the treatment the alloys were cut at 5 cm distance of the crease point to avoid the curly ends.

First experiments were carried out for 5 min, at temperatures of 500 and 550 °C. New experiments were developed at a temperature of 550 °C and varying the period of time for each sample set—20, 15 and 10 min—a sample set being composed by five Nitinol samples of each of the three wire diameters, totaling fifteen.

The prepared samples were then analyzed by submitting the wires to five cycles of deformation and heat stimulus. For this purpose, they were placed one at a time in a heat chamber, after being deformed at the cooling phase—martensite (M). In the chamber, temperature was increased from 20 to 45 °C (Af temperature) and, for each cycle, the sample was removed, placed on a surface with a guideline and photographed. The angles attained on the captured images were measured in Adobe Illustrator software.

The angles measured above 95° in the samples that were heat treated during 5 and 10 min. At 20 min, the wires presented a significant variation of angles between the different cycles. In the wires treated during 15 min, angles between 90° and 92° for Nitinol of 0.2 mm thickness (1st row of Fig. 6.7); 90° to 91° for Nitinol of 0.3 mm (2nd row of Fig. 6.7) and of 0.5 mm (3rd row of Fig. 6.7) were obtained.

Considering that the results attained with the heat treatment of 550 °C during 15 min present small, acceptable variations from the pre-programmed 90° angle, this condition was selected for further experimentation.

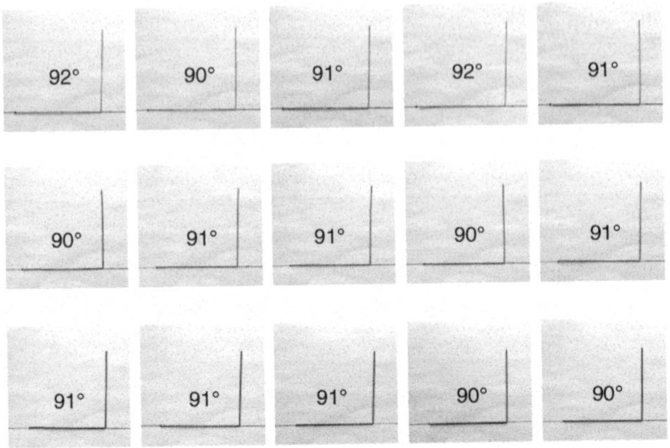

Fig. 6.7 Angle values attained throughout 5 cycle tests for each Nitinol wire, with heat treatment during 15 min. Upper row to lower row: 0.2; 0.3 and 0.5 mm diameter (Cabral et al. 2015)

6.4 Textile One-Way Shape Memory Behavior

The objective of this experiment was to study the integration process of Nitinol wires in woven substrates and their ability to recover the defined morphology when activated. The predefined geometry of the shape memory textiles dictates the Nitinol wires' shapes, according to the crease pattern and the specific location where each alloy will be integrated. This condition highlights the drawing and manufacturing of different wire geometries to be embedded in the same textile structure, as it will be illustrated.

In order to study the shape change behavior within predefined geometric configurations, the origami base folds—waterbomb and squaretwist—were implemented on basis of woven samples, by integrating Nitinol wires of the three selected diameters.

The integration of Nitinol wires was done in a table handloom, with a plain weave structure. This selection was due to the regularity and maximum number of interlacings of this woven structure and their improved control of Nitinol for the shape change (Dyer 2010). It was defined that the final samples dimensions would be 12 × 12 cm with Nitinol wires inserted in the weft at every 1.5 cm.

To achieve the final folding effect, the shapes to be memorized on each of the Nitinol wires depend on the origami morphology while folded. To define these wire geometries, a paper model with horizontal lines spaced by 1.5 cm was folded with the shape of the selected origami. The resulting wire geometries were then taken from the model. Each model includes seven Nitinol wires and some of the wire configurations are equal. The waterbomb model contains four different wire shapes (Fig. 6.8), whilst the squaretwist model has three (Fig. 6.9).

After defining the Nitinol wire shapes, the die for each geometry was designed and produced (Fig. 6.10). The wires' extremities were fixed with screws and the wire shape followed the die geometry by passing orifices placed at the exact crease locations.

Fig. 6.8 Waterbomb model open and close (left and center); SMA shapes (right) (Cabral et al. 2015)

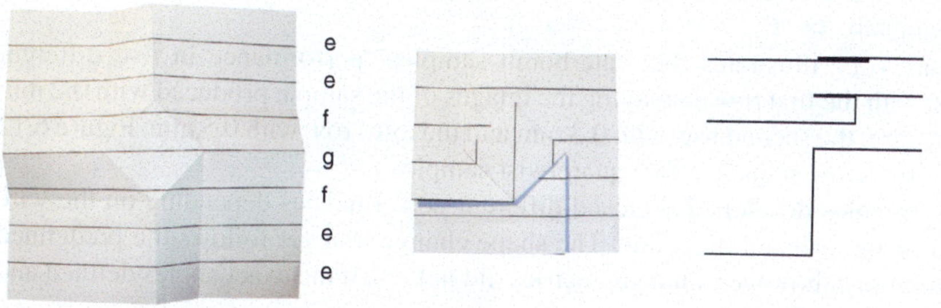

Fig. 6.9 Squaretwist model open and close (left and center); SMA shapes (right) (Cabral et al. 2015)

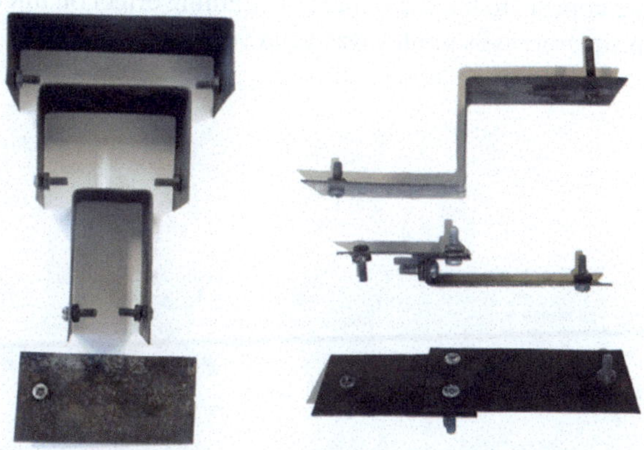

Fig. 6.10 Waterbomb dies (left); Squaretwist dies (right) (Cabral et al. 2015)

Finally, samples of the base structures were produced, which involves the following steps: training the Nitinol wires with the predefined shapes; setting up the loom and weaving the samples. A total of six samples were developed—for each origami model (waterbomb and squaretwist) three samples were produced with the different SMA wire thickness (0.2; 0.3 and 0.5 mm).

Nitinol training consisted of fixing the wire, one at the time, in each die and placing it in the furnace for 15 min, which was pre-heated at 550 °C. When removed from the furnace, samples were water quenched, being exposed to a temperature of approximately 10 °C and were then deformed to a straight shape.

The samples were woven in a table handloom. The same CO yarn was used for the warp and the weft and one specific pre-trained Nitinol wire was inserted every 1.5 cm, according to its position in the origami crease pattern, into a plain weave structure.

To analyze the behavior of the developed study samples, it was conducted a video and photographic record. The samples were deformed at M phase, placed on a horizontal surface and thermal activation was induced with a heater, attaining approximately 45 °C.

Figure 6.11 illustrates the waterbomb samples' performance in five different phases, with the first row displaying the images of the sample produced with 0.2 mm Nitinol wire; the second row with 0.3 mm and the third row with 0.5 mm. Figure 6.12 follows the same sequence for squaretwist samples.

The samples developed achieved different performances depending on the wire thickness and origami diagrams. The shape change transfer follows the predefined configurations, however, final geometries did not reach the exact angles defined and trained in the Nitinol wires.

In the waterbomb configuration, the sample with 0.5 mm Nitinol exhibits better shape recovery than the ones of 0.3 and 0.2 mm. Conversely, in the squaretwist origami, the 0.5 mm sample achieved the lowest performance. Due to the existence of more creases in the squaretwist diagram than in the waterbomb one, the woven structure appears to have a higher constraining effect on thicker wires. Better performance on squaretwist samples was achieved with 0.3 and 0.2 mm Nitinol.

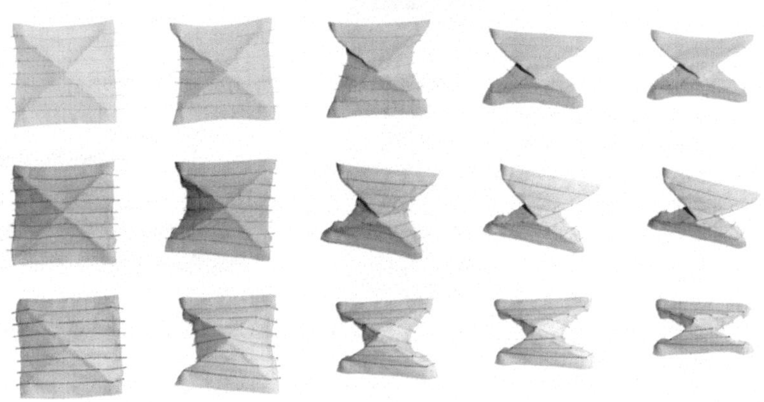

Fig. 6.11 Waterbomb samples during thermal activation (1st row 0.2 mm wire sample, 2nd 0.3 mm and 3rd 0.5 mm) (Cabral et al. 2015)

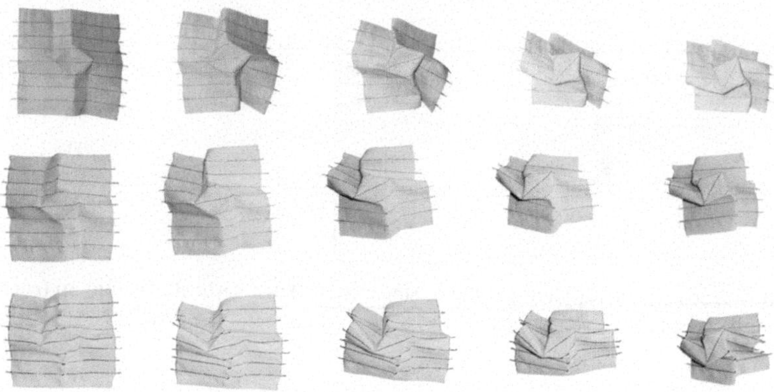

Fig. 6.12 Squaretwist samples during thermal activation (1st row 0.2 mm wire sample; 2nd 0.3 mm and 3rd 0.5 mm) (Cabral et al. 2015)

The shape change process was recorded in three videos for each sample. In both origamis, the samples with 0.2 mm Nitinol wires needed more time to recover their shape when compared with other samples, of approximately 50 s. In the waterbomb diagram, the average duration for samples developed with 0.3 and 0.5 mm was of 18 s, whilst in the squaretwist diagram, the values were slightly higher: 22 s for 0.3 mm and 25 s for 0.5 mm samples.

Difficulties were found in deforming the samples on flat surfaces. It was not possible to deform the Nitinol creases to assume a straight shape by hand deformation. In some of the creases, wire protruding the woven structure was observed. This was more evident in samples with 0.5 mm thick wire.

Nitinol 0.3 mm thickness wire was selected for the following developments, given that its overall behavior was found more stable in regards to the aforementioned variables. Furthermore, weave density is to be increased to improve shape transfer of the Nitinol integrated in the textile.

The second phase of this study applies the process setup proposed in the preliminary experiments, to the design of a shape memory textile based on an intricate origami. The sample development encompasses an increase of the weave density with the aim to improve shape transfer of the Nitinol integrated in the textile. The shape was defined with basis on the geometry of squaretwist tabs fold up and the sample activation was triggered through resistive heating. Finally, light transmittance variation generated through the sample performance was analyzed.

To define the textile morphology, the base fold selected was the squaretwist. In the folded state, this origami presents a square area with five layers, four tabs with three layers (in the area exterior to the square) and a remaining area with one. The tabs are two by two, parallel to the correspondent square diagonals (vertical and horizontal). When tabs with the same direction are pulled up to an orthogonal plane, in relation to the folded origami, it is possible to obtain a valley fold in a square diagonal attaining a variation in the amount of layers. When the square and two tabs are folded up,

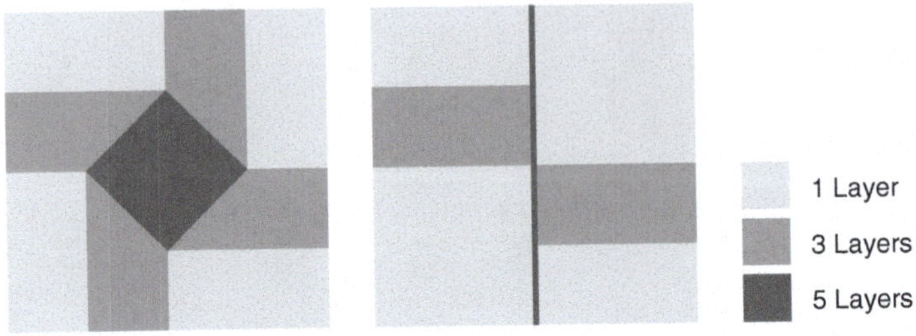

Fig. 6.13 Layer number variation per areas: squaretwist (left), vertical tabs fold up (center) and color subtitles (right)

what remains in the origami surface are two tabs with three layers and one layer in the overall structure, as illustrated in Fig. 6.13.

The shape change described was implemented in the sample development. The origami crease pattern designed combines eight squaretwists with asymmetrical distribution in a quadrangular frame. Dash-dot green lines in the crease pattern presented in Fig. 6.14, depict Nitinol wires placement. Seventeen Nitinol segments were defined.

As previously explained for the base folds study, once again the origami creases define the geometry that the wires should be trained with. After folding the model with the configuration that should be assumed, when heated, each wire shape was drawn (Fig. 6.15 left) and subsequently, the die for each shape was drawn and produced (Fig. 6.15 right). This model displays five different wire shapes. Numbered from top

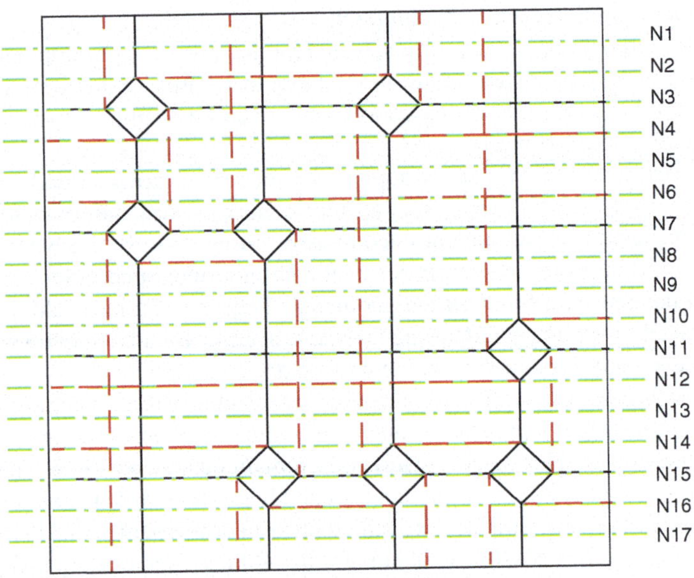

Fig. 6.14 Sample crease pattern (Cabral et al. 2015)

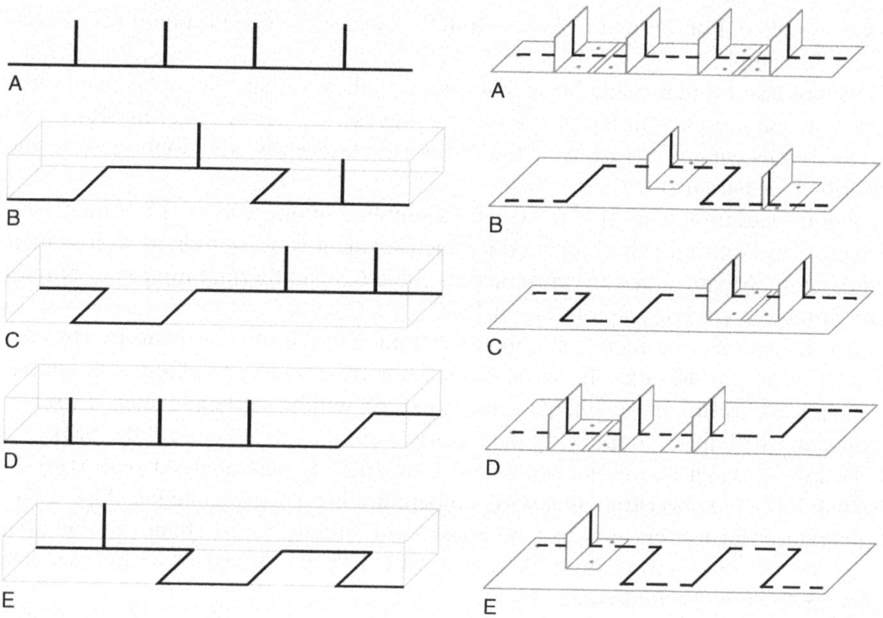

Fig. 6.15 Nitinol wires' shape (left); Nitinol dies' shape (right) (Cabral et al. 2015)

to bottom, *A* corresponds to wire number 1, 2, 4, 5, 6, 8, 9, 10, 12, 13, 14, 16, 17; *B* is 3; *C* is 7; *D* is 11 and *E* is 15.

Nitinol size and the mass of the die can influence the optimization of the temperature and time parameters of the annealing process (Liu et al. 2008; Rao et al. 2015), they were retested and the values set were 550 °C for 10 min.

The weaving process was developed in the Jacquard loom with the first set up (Fig. 6.1). Nitinol segments have been integrated manually in order to assure accuracy within the crease pattern. The alloys were deformed to a straight shape at M temperature and inserted in each section of 21 textile picks, totalizing 2 cm distance between each Nitinol wire. To maintain a straight shape in the correct location, the wire ends were held during the next weft textile yarn insertion.

The thermal activation of the shape memory textile was produced through resistive heating—a method commonly applied to heat shape memory textiles (Lee 2005; Berzowska and Coelho 2005; Seymour 2008, 2010; Coelho and Zigelbaum 2011).

The electrical current required to attain the Nitinol activation temperature depends on the Nitinol characteristics, room temperature and duration of power supply. A study to specify the current value was conducted, to ensure that the Af temperature would be attained (45 °C) without overheating the wire.

The experiment took place at a room temperature of 23.8 °C. For each Nitinol segment integrated in the woven structure, a specific voltage was applied. The resulting electrical current was measured and a thermal image was recorded, after 5 s of power supplied. For thermal isolation from the table surface, a PE board was placed below the sample.

Nitinol has a tough oxide layer that does not allow an accurate resistance measurement and presents difficulties in joining processes, such as soldering (Rao et al. 2015). A mechanical abrasion of the Nitinol surface ends was done prior to the described measurement.

Voltage variation from 0.5 to 3.0 V was supplied in intervals of 0.5 V increases. The electrical current values attained for each segment were similar, as well as their temperature. Mean values are presented in Table 6.3 and thermal images of Nitinol wire number 3 are presented in Fig. 6.16.

An analogous experiment was conducted linking the Nitinol segments two by two, in a series circuit. Relation between electrical current and temperature was similar to the results within one segment, as expected. As will be explained later, the series connection will allow simplifying the electrical circuit used to control the SME.

Despite a mean maximum temperature of 46.8 °C was attained with 0.63 A, thermal images demonstrate that wire temperature is not homogeneous. Also, room temperature during the tests can be considered slightly higher than the common definition of room temperature. A current of 0.7 A was found to be an adequate value to drive the Nitinol wires.

In order to achieve a controlled heating of the Nitinol wires, and considering that the simultaneous activation of all 17 wires would require a considerable amount

Table 6.3 Mean values of electrical current and maximum temperatures attained

V (V)	I (A)	Temp. (°C)
0.50	0.12	26.20
1.00	0.23	27.80
1.50	0.37	32.20
2.00	0.51	37.80
2.50	0.63	46.80
3.00	0.75	54.10

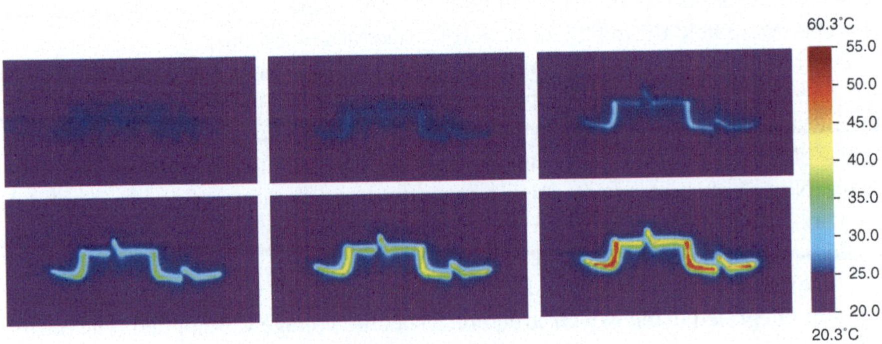

Fig. 6.16 Thermal images of Nitinol wire number 3 during voltage variation

of current, a sequential activation was implemented. For this purpose, an Arduino microcontroller board was used to control the process. The power stage was implemented using ULN2003A Darlington-transistor arrays, which feature 7 channels with a maximum current of 0.5 A each. To drive the wires with 0.7 A, it was necessary to use two channels in parallel for each wire, reducing the number of controllable wires to 3 per each transistor. To simplify the circuit, wire segments on the textile were paired and connected in series, which allows reducing the number of wires to control to 8 + 1 (8 pairs and one single), using three ULN2003A chips instead of 6. Figure 6.17 shows the circuit diagram of the power stage. The Nitinol wires are connected in series with pull-up resistor R_p, which has been dimensioned to allow a current of 0.7 A. All of the resistors are of equal value except the one which drives the single wire. Figure 6.18 shows the very simple code used to activate the wires sequentially, with a heating time of 6 s and a delay of 1 s between channels.

Having completed the prototype development, an analysis focused on shape change performance, thermal measurement and transmittance variation was conducted. The shape change performance was recorded on video and the frame of the final textile morphology attained was compared with the predefined geometry. A thermal imaging record was made to analyze temperature in each Nitinol wire during current flow. Light transmittance variation was measured before and after activation (temperature < As and temperature > As, respectively), with a similar setup described for the results presented previously in Table 6.2.

The photographic record of the prototype, before and after activation, is presented in Fig. 6.19. For a visual comparison of the morphology achieved and defined, an image of the paper model depicting the defined activated morphology was also included.

It is possible to conclude that some areas achieved equal or similar configurations to the predefined ones. There is a main incongruity in the bottom right square area, as the tabs are not perpendicular to the prototype surface, resulting in a non-folded square.

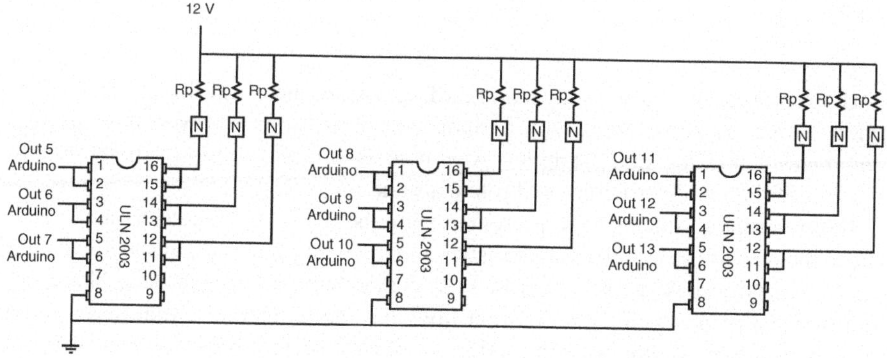

Fig. 6.17 Circuit diagram with "N" representing the Nitinol wires (Cabral et al. 2015)

```
/*
  */

// These constants won't change.  They're used to give names
// to the pins used:
const int digitalOutPins[9] = {13,12,11,10,9,8,7,6,5};//

void setup() {
  // initialize serial communications at 9600 bps:
 for (int i=0; i <= 8; i++)
   pinMode(digitalOutPins[i], OUTPUT);
   delay(3000);
   for (int i=0; i <= 8; i++){
  digitalWrite(digitalOutPins[i], HIGH);
  delay(6000);
  digitalWrite(digitalOutPins[i], LOW);
  delay(1000);
  }

}

 void loop() {
 delay(100);
   }
```

Fig. 6.18 Arduino code (Cabral et al. 2015)

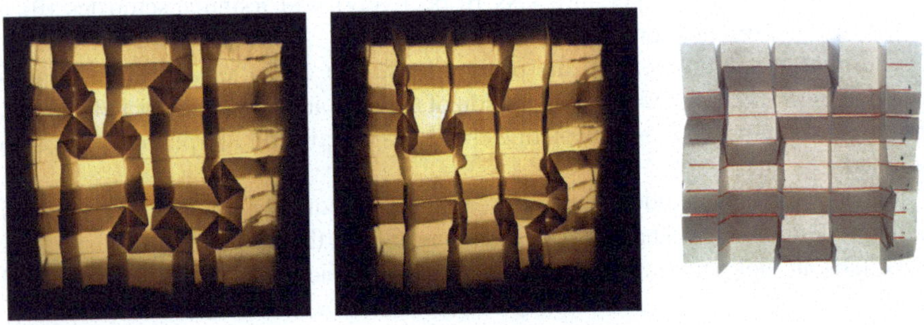

Fig. 6.19 Prototype before activation (left); activated (middle) and paper model (right) (Cabral et al. 2015)

The prototype was also analyzed regarding textile and Nitinol materials' interaction. There are some areas with Nitinol wire protrusion; however, they are less prominent than in the study samples. Also perceived is the ability of the Nitinol to slide out of the textile structure and rotate around its axis.

Thermographic images have presented some issues on the electrical activation. Due to the origami structure characteristics and the wire protrusion in specific crease areas, some Nitinol segments are close or in contact with other segments. This compromised the planned electrical current flow and the temperature that needs to be attained. Figure 6.20 illustrates the activation of Nitinol wires number 9 and 10. It is possible to detect a short circuit area, in which there is an overheating, as defined by

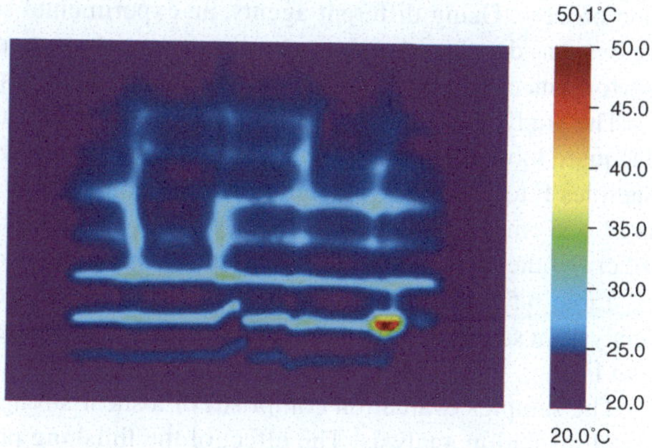

Fig. 6.20 Thermal images of Nitinol number 9 and 10, being activated (Cabral et al. 2015)

Table 6.4 Mean values of light intensity

Light intensity (lx)		Luminosity difference (%)
Temp. < As	Temp. > Af	
0.56	0.97	42.27

the dark red area, being the remaining wire below the activation temperature. This explains the described anomaly in the bottom right area of the textile.

The aforementioned issues are due to the smoothness of the Nitinol surface and the fact that they are not electrically isolated. This finding has delineated the development of a future study with the purpose of implementing a finishing process in order to increase adhesion and thus avoiding slip, rotation and protrusion of the Nitinol from the textile substrate; to isolate the wire to avoid issues with undesired electrical contact between wires (Cabral et al. 2015).

The origami scale defined also appears to be relatively small in comparison to the textile's thickness. In some areas, the overlapped layers constrain the SME, reducing the ability to perform with greater accuracy within the predefined geometry.

During the observation and recording of the shape change, transmittance variation was perceptible. The light intensity measurement provided a quantification of the variation. The luxmeter was pointed to a white surface placed 1 m from the light box and the values were recorded on three points. Mean values reached a percentage difference of 42% (Table 6.4).

6.5 Finishing Processes

Nitinol surface can be modified in order to customize their qualities to the applications requirements (Shabalovskaya et al. 2008). In this study, finishing processes were applied in woven textiles that integrate Nitinol alloys, aiming to improve adhesion of the Nitinol with the host structure and electrical insulation of the alloys in one process. Using different agents, an experimental study of textile finishing processes was developed and a comparative analysis is reported, regarding mechanical performance, electrical insulation, weight and thickness of the developed samples.

The study of the finishing processes was conducted with samples woven in the Jacquard loom Bonas. Samples A to E were produced with the first setup of the loom, samples F to I were developed with the second set up (Fig. 6.1). The samples' size is 11 × 5 cm with one Nitinol wire previous annealed with a 90° angle, integrated in 10 cm of the weft. Figure 6.21 depicts the samples' geometry and dimensions (cm).

Table 6.5 presents a description of the samples developed. For a comparative analysis, a sample without treatment of each loom set up was developed (samples A and F).

The samples evaluation comprises of a shear strength measurement and an electrical insulation analysis. The effect of the finishing processes in the Nitinol ability to slide of the textile substrate was measured in a dynamometer with a breaking strength and elongation test. The sample was placed in the dynamometer with the Nitinol end attached to the superior grip and the opposite sample side was hold in the textile area with the lower grip. To analyze if the electrical insulation was attained, a multimeter continuity test was conducted with one test lead connected to the Nitinol end and the other touching the Nitinol crease point.

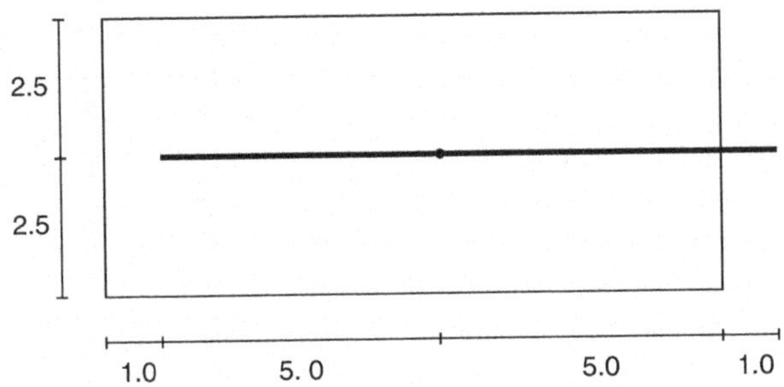

Fig. 6.21 Schematic of the samples' geometry

Table 6.5 Description of samples' materials and finishing processes

Samples	Material no.	Finishing process
A	–	–
B	1	Screen printing
C	2	Screen printing
D	1	Coating
E	2	Coating
F	–	–
G	3	Impregnation
H	4	Impregnation
I	5	Spraying

Additionally, has the textile thickness and weight also can influence the shape memory textile behavior, these samples' properties were also measured and compared.

For a greater understanding of the color and shape change possibilities, after the selection of one of the material and finishing process, a screen printed sample was developed with a 10% TC red paste, before the finishing application and evaluated within the aforementioned parameters.

Table 6.6 presents the result of the breaking strength and elongation tests. Comparing A and E samples, it is possible to verify that the textile substrate characteristics of A present higher interfacial adhesion of Nitinol alloys with the host structure than E. In relation to the processes and materials applied, screen printed samples attained a greater increase of the force necessary to initiate the removal of the alloy from the woven structure and the value is higher for sample B. Samples developed with impregnation also attained relevant increase of adhesion while pulverization attained low and samples with knife-coating process presented values approximating zero and D testing was inconclusive.

Regarding the multimeter continuity test, just sample I have presented electrical insulation of the Nitinol alloy. The remaining samples have maintained conductivity behavior (Cabral and Souto 2015).

Samples thickness and weight values are presented in Table 6.7. Sample A, developed in first loom setup, presents thicker and heavier characteristics than sample F, woven with the second setup. Impregnation process in samples G and H, attained the higher increase of the measured characteristics and the pulverization (sample I) presents the lowest increase.

Sample I has a low adhesion increase value in comparison with screen printed and impregnated samples. However, it was the solution that had provided electrical insulation and presents the lowest increase of thickness and weight values. Thus, it was the option selected to apply in the following samples development.

The color and shape change sample developed attained 2.85 N with 0.15% extension on the breaking strength and elongation test, it is electrically insulated in the Nitinol integration area and the thickness and weight values are 0.52 mm and 0.848 g, respectively.

Textile structure characteristics play a significant role in Nitinol ability to perform the shape change transfer to the host structure (Dyer 2010; Vasile et al. 2010). With the objective to study the effect of the finishing process selected in the shape change behavior, a comparative analysis was conducted to observe the shape change of samples with and without the finishing process. With the perspective to develop dynamic light filters that perform color and shape change behavior, samples were also screen printed with TC pigments.

The first test was conducted with 4 samples, two were woven in the electronic Jacquard loom with PES warp and the other two in Jacquard Vamatex loom with the coarser warp setup. They have followed the same geometry depicted in Fig. 6.21. One sample of each loom type was screen printed with a 10% TC red paste, thermo-fixed and later the finishing agent was applied. The remaining samples were not treated.

Table 6.6 Breaking strength and elongation values

Sample	A	B	C	D	E	F	G	H	I
Breaking strength (N)	0.65	31.65	27.65	–	0.35	0.35	6.15	9.50	0.85
Extension (%)	0.50	7.73	5.00	–	0.60	0.50	0.85	1.60	0.15

Table 6.7 Samples thickness and weight

Sample	A	B	C	D	E	F	G	H	I
Thickness (mm)	0.47	0.49	0.49	0.51	0.51	0.41	0.47	0.63	0.45
Weight (g)	0.86	1.13	1.4	1.02	1.15	0.63	2.72	1.76	0.69

To study the shape change behavior, the samples were hanged with the Nitinol wire deformed in straight shape and placed in a vertical position. A power supply plug was directly linked to the upper end of the alloy, the lower end was connected to a conductive thread with a crimp bead and linked to the power supply. For each sample, it was driven an electrical current of 0.7 A. The samples' behavior was recorded on video and the frames corresponding to the shape before and after activation stage were compared, between each group of samples.

With the objective to study the finishing effect in a more complex structure, a sample with squaretwists pattern was also analyzed. The crease pattern and the shape change definition of the sample corresponds to the lower area of the model developed in Sect. 6.4, from Nitinol number 9 to number 17 (Fig. 6.14). This sample was woven in the Jacquard Vamatex loom with the coarser CO warp and each Nitinol wire was integrated after 27 textile picks. In this sample, the Nitinol alloys were linked in 2 series circuits, the sample was folded, placed in horizontal position and their behavior was video recorded. After, the sample was unfolded, screen printed with a 8% TC red and 2% TC black paste, thermo fixed, the finishing agent was applied and the aforementioned analysis process was repeated.

The activation temperature of the Nitinol handled (Af 45 °C) was higher than the TC pigments selected (27 °C), thus, the shape change behavior will occur simultaneously to the color change, revealing the Nitinol integration pattern in the textile color.

The images captured on the first test are presented in Fig. 6.22. By a visual analysis it is possible to identify that after activation, samples not treated attained an angle more similar to Nitinol 90° geometry than samples treated. Comparing the samples woven equipment, Bonas loom (loom A) has a lower number of interlacings and

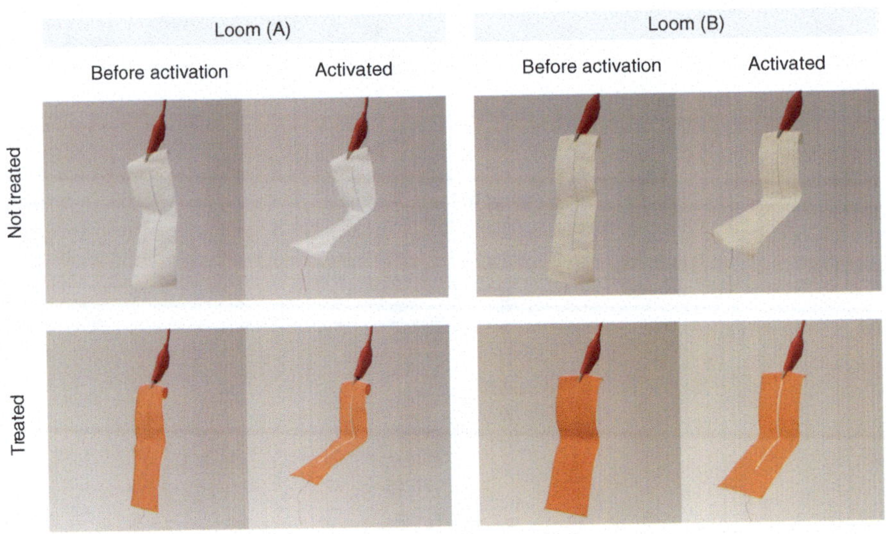

Fig. 6.22 Samples before activation and activated

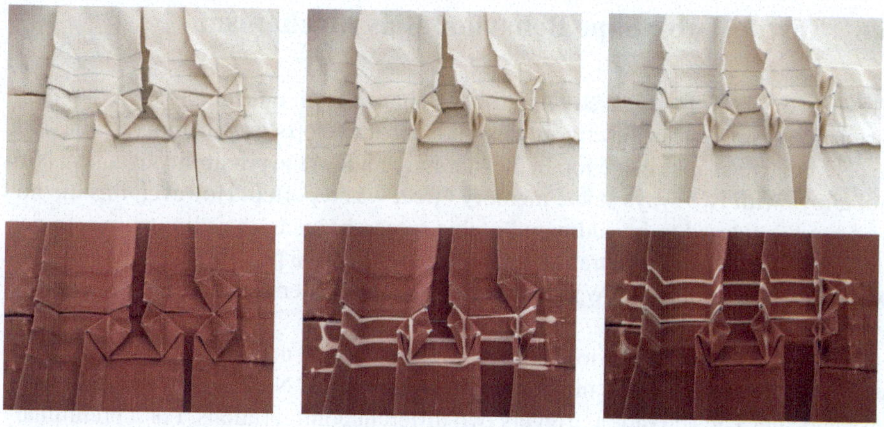

Fig. 6.23 Sample before and after treatment during sequential activation

samples are lighter than in Vamatex loom (loom B) and have performed a higher angle of shape change.

The origami sample was activated in two sequences, starting in the lower area and after in the upper area. The frames selected to conduct this analysis refer to the sample deformed at M temperature, activated in sequence 1 and 2. First and second raw of Fig. 6.23 illustrates the sample behavior before and after treatment, respectively.

Before the finishing processes, the sample attained the fold up squares more near to the orthogonal position than after treatment. The tabs show similar shape in the areas with Nitinol wires integration. However, the sample tops do not integrate the alloys and before treatment the textile is softer and do not follow the shape change movement as well as in the sample treated.

In both activation tests, it was observed that the shape change is not similar in the overall sample. The squares located in the right side were able to perform a higher fold up angle than the other squares. This heterogeneous behavior appears to be related to textile contraction after screen printing, wire protrusion in some points and layer composition along the substrate thickness, affecting the accuracy of the Nitinol wires geometry in regards to the folded host structure and resulting in areas with different SME constrains.

This study attained a visual understanding of the effect of selected finishing process in the shape memory behavior and the possibility to combine shape change and color change in the same textile. The results also highlight the need to study alternative approaches of how to integrate Nitinol wires in woven structures, towards higher similarities of shape change attained in comparison to the predefined one. This requirement will be further researched in the next section.

6.6 Textile Two Shape Behavior: Bias Mechanism

Previous experiments focused on the shape change from one temporary to a memorized morphology. The objective of this study is to develop shape memory textiles able to perform two shapes behavior. In the literature review, the possibility to anneal Nitinol alloys with two-ways shape memory effect was reported. However, this method was not considered to be addressed in this research due to their limitations, in particular the decrease of the transformation forces and the possibility of memory lost with a slight overheating (Wan and Stylios 2007; Rottiers et al. 2011; Bengisu and Ferrara 2018).

The shape change behavior between two memorized morphologies was explored with a bias mechanism, the integration of two groups of Nitinol alloys whose geometry and alternate activation, triggers two different configurations. For a preliminary analysis, the two shapes behavior was researched in a flat study sample.

The sample was woven in the electronic Jacquard loom with PES warp, integrating six Nitinol wires: three with a 90° angle and the other three annealed with a straight shape (Fig. 6.24 light and dark green lines, respectively). The Nitinol wires defined to perform each shape were called actuator groups and the alloys were integrated in intercalated arrangement at every 2 cm in 10 cm of the weft.

To analyze shape change behavior, the alloys of each actuator group were connected in a series circuit and activated in different moments. First, the sample was deformed to a flat shape at M temperature and the 90° angle alloys group was driven with 0.7 A during 10 s followed by a pause of 5 s. After, straight shape actuators were activated. The sample behavior was recorded on video and the frames corresponding to the described activation stages were analyzed.

Figure 6.25 presents the sample deformed into a flat shape (left image), first actuator group activated (center) and second actuator group activated (right).

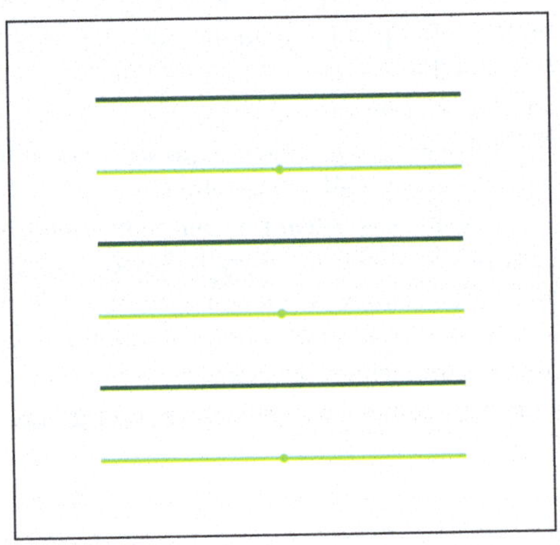

Fig. 6.24 Diagram of Nitinol alloys configuration

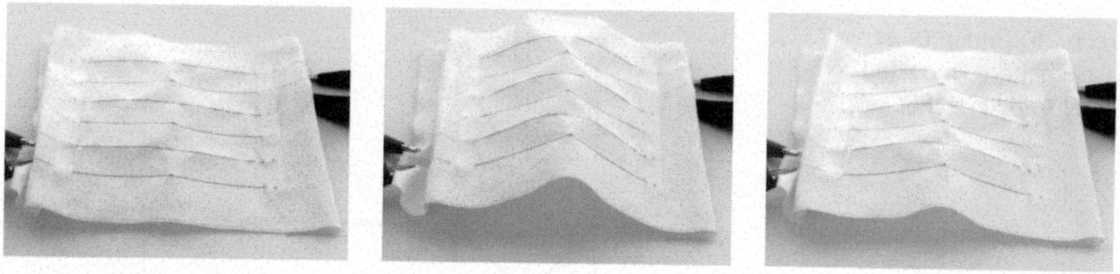

Fig. 6.25 Sample deformed (left); 1st Actuator group activated (center) and 2nd Actuator group activated (right)

The image on the left presents an overall flat shape however, as the Nitinol creases are not able to be completely deformed, they create a small undulation on the surface. The activation of the 90° actuator group have performed the shape change of the textile while also deforming the second actuator group. The non-activated alloys present forces contrary to the movement of the activated group influencing the predefined angle attained, which did not reach the 90° shape. The same effect happens during activation of the straight shape group. Comparing the sample manually deformed in flat shape and the sample with the second actuator group activated, they should present similar shapes but the activated result was not as flat as the predefined shape.

Despite the morphology accuracy was lower than the pre-defined, the shape change was evident and the sample attained two different morphologies. The proposed bias mechanism was found valid to be implemented and further researched in self-folding textiles.

In order to further research the two shapes behavior in origami structures and address the shape accuracy improvement aim, a different Nitinol integration approach was proposed. Nitinol wire integration across the full width of the host structure enhances the influence of the textile substrate constrains in the SME, across the Nitinol wire protrusion and geometry accuracy in regards to the folded structure, as discussed in Sect. 6.5. Furthermore, some sections of the alloys do not convey shape change, their purpose is to connect the actuator parts along the samples' width. This connection can be developed with conductive threads, allowing greater flexibility to define the Nitinol diagrams and maintaining the required conductivity between the alloys, a method approached in the following sections.

In this study, Nitinol integration in the woven substrate will be defined by segments, in specific areas of the origami crease pattern, taking into consideration the crease axis sections. Previous to the sample development, the textile contraction attained after screen printing and thermo setting processes was also analyzed, to scale up the weft density of the woven substrate in order to increase the morphology accuracy.

To fold up and fold down the vertical tabs and square area of the squaretwist origami, the main crease axis corresponds to the vertical valley folds and the horizontal diagonal of the square (that when is folded, it twists and becomes vertical). The grey shadows in Fig. 6.26 (left) highlights these creases axis in the unfolded

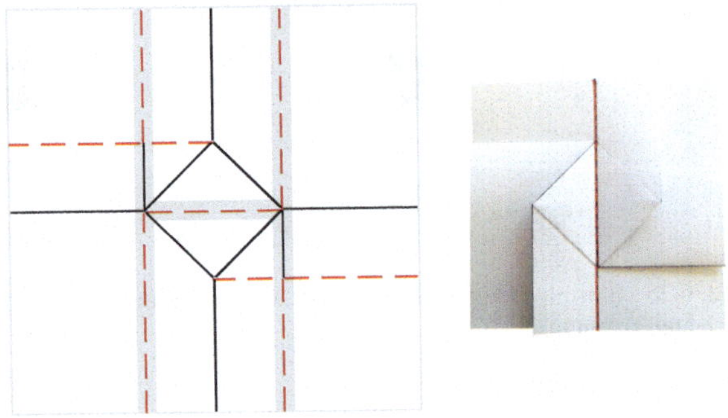

Fig. 6.26 Squaretwist crease pattern (left) and model folded (right)

origami and when the diagram is folded, they overlap in one central axis (vertical red line).

The models' layers and overlapped creases were studied in order to define the main axis for Nitinol integration areas. Crease pattern of samples with two Nitinol actuation groups will be depicted with black continuous line for mountain fold and black dashed line for valley fold. The actuator group performing fold up behavior is illustrated with light green lines and the fold down in dark green. The points added in the Nitinol lines identify the alloy's crease points.

The samples' shape defined for this study consists of squaretwist fold with 6 cm square diagonal. In sample S6 A (Squaretwist 6 cm), the Nitinol creases are aligned in the vertical valley fold corresponding to the lower layers of the folded model. The alloys of the two actuator groups, present four different geometries (Fig. 6.27).

Sample S6 B integrates the alloys in the crease axis of the origami upper layer and also present four geometries. In this case, the fold down actuator group (dark

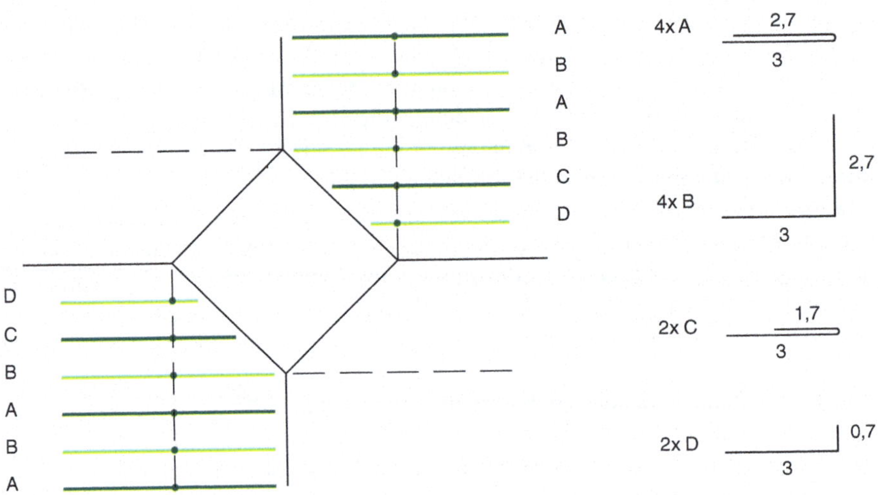

Fig. 6.27 Sample S6 A crease pattern (left) and Nitinol wire geometries (right)

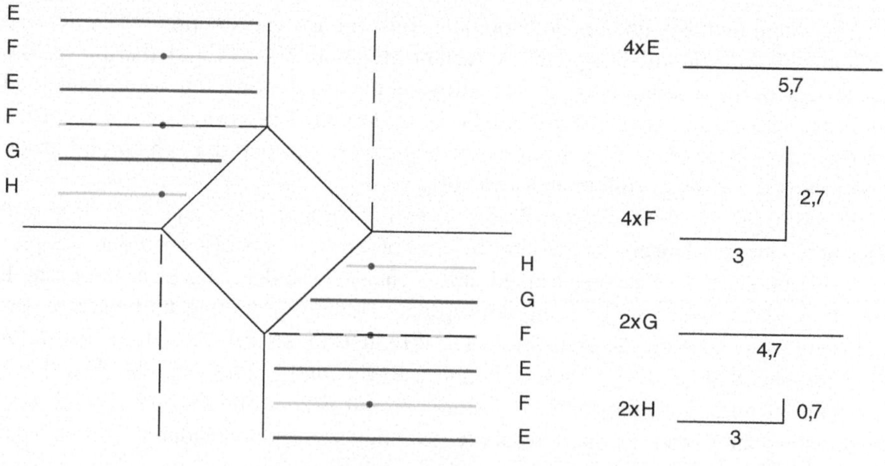

Fig. 6.28 Sample S6 B crease pattern (left) and Nitinol geometries (right)

green lines E and G of Fig. 6.28), instead of performing the decrease of a 90° angle as in S6 A, they have to increase it until 180°.

A 3rd sample (S6 C) was developed combining the previous samples' actuators geometries. In the areas exterior to the square, the Nitinol alloys are integrated in the full width of the sample (Fig. 6.29).

The samples were woven in the electronic Jacquard loom with PES warp and 1 cm of Nitinol spacing. Previous to their development, a test sample was produced in the same equipment with Nitinol wires integrated at every 1 cm, in order to analyze textile contraction after being screen printed and thermo set. The contraction value was approximately 10%, establishing an increase of 1 pick/cm in relation to previous samples, thus 12 textile picks plus 1Nitinol wire per cm.

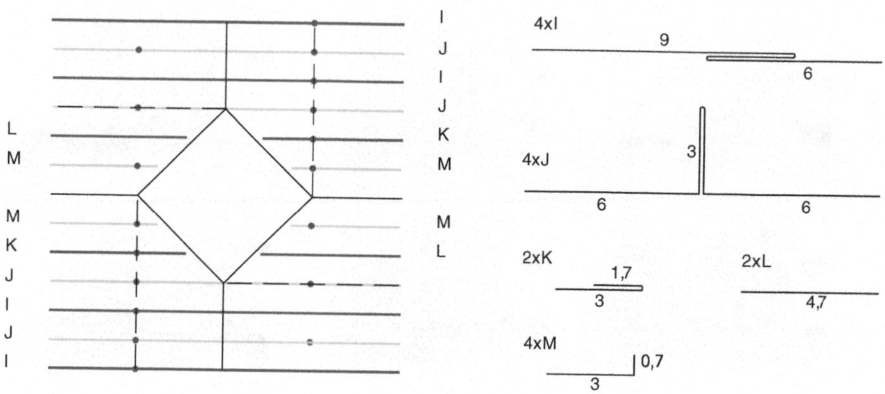

Fig. 6.29 Sample S6 C crease pattern (left) and Nitinol geometries (right)

The shape memory textiles development followed the process previously defined in Sect. 6.4. Additionally, they were screen printed with 10% TC red, thermo set and finished with the insulation agent. The alloys of each actuator group were connected in a series circuit and, after the origami being folded, their performance was recorded on video and the frames of each activation stage were compared. Each Nitinol group was driven 0.7 A by a Velleman power supply.

Figure 6.30 presents the samples' images before activation and at fold up and fold down stages. During the manual folding process, it was observed that samples A and B presented a less constrained ability than C, the decrease of Nitinol length improved the textile handling and appearance. However, the fold up behavior was not compromised with the actuator material reduction, sample A attained a similar fold up square than C, but sample B had a lower angle behavior. The fold down was similar among all the samples. Sample A and B also present less Nitinol wire protrusion than C and, as the Nitinol creases happen in the considered back side of the samples A and B, they are not visually perceived.

The development of two shapes behavior in origami textiles with the proposed approach of Nitinol integration based in the main creases presents important advantages in comparison to the Nitinol integration in the full samples' width, particularly within sample S6 A setup. The sample presents greater folding accuracy within less Nitinol material and thus less expensive. The decrease of Nitinol crease points reduces the protrusion number and they are less prominent. Furthermore, it is also

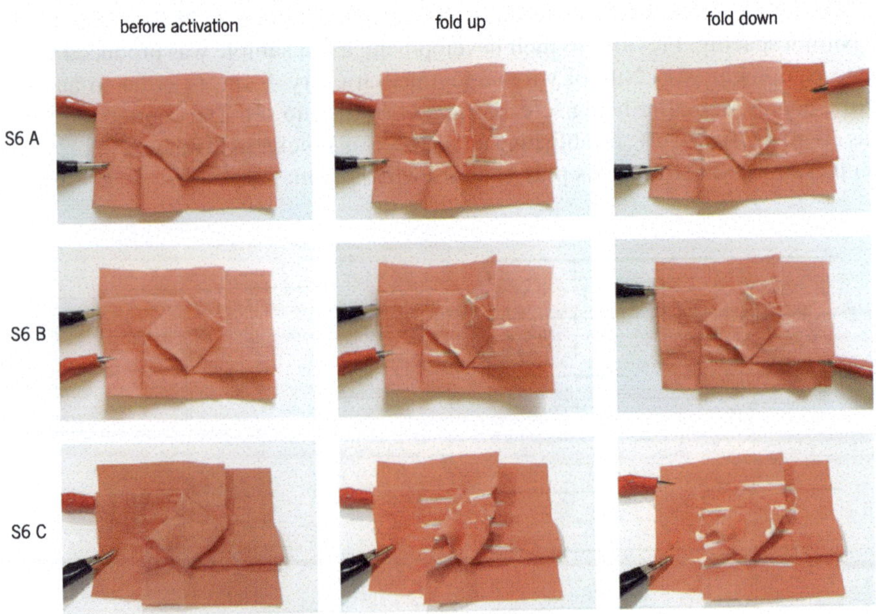

Fig. 6.30 Samples behavior comparison: S6A, S6B and S6C

easier and faster to develop, the dies are simpler and they have a lower number of segments to insert in the woven structure.

The similarities between the shapes defined and attained still demands experimentation. The proposed approach of Nitinol segments in specific areas of the textile, allows the possibility to increase the memorized Nitinol angle to compensate the contrary forces to the movement. It was selected the sample A main crease axis definition for the Nitinol integration, to be implemented in further experimental work.

The previous samples developed attained shape change behavior with lower angles than the geometries annealed in the Nitinol wires. In order to compensate the SME constraints in the alloys shape transfer, the follow study proposes to increase the Nitinol angles definition in relation to the intended textile performance.

Plain and squaretwist samples were developed to study the effect of the Nitinol memorized angle in the textiles shape change. All samples were woven in the electronic Jacquard loom with PES warp.

Plain samples are composed by two Nitinol wires with 10 cm integrated in the weft—one alloy performs the fold up and the other, the fold down. With the objective to attain a fold up behavior of 90°, the alloys' angles defined for this behavior were 100°, 110° and 120°, comprising two different spacing between the fold down alloys—1 and 2 cm—totaling 6 samples. Nitinol fold down angle it is equal to the S6 A samples, 0°.

The increase of the fold up wire's angle was expected to affect the fold down. With the squaretwist model, two samples were developed with 110° fold up actuators with different integration setups: S6 D sample presents a regular Nitinol spacing intercalating one fold up with one fold down actuator (Fig. 6.31 left), S6 E has two fold down alloys at 0.5 cm distance, per each fold up at 0.75 cm (Fig. 6.31 right). After their development and performance analysis, two additional samples were produced following the S6 E setup, with a higher angle definition: S6 F present 120° angle of the fold up alloys and S6 G have 130°. The samples development and analysis followed the process described in the previously described.

Figure 6.32 shows the results attained with the plain samples. Comparing the fold

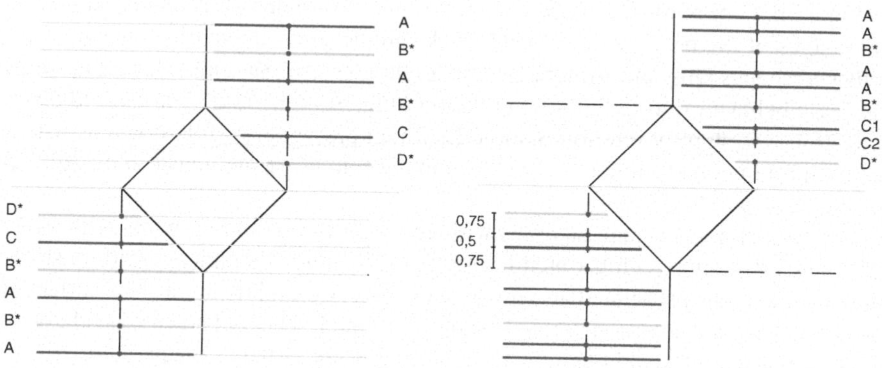

Fig. 6.31 Crease pattern samples S6 D, F and G (left), S6 E (right)

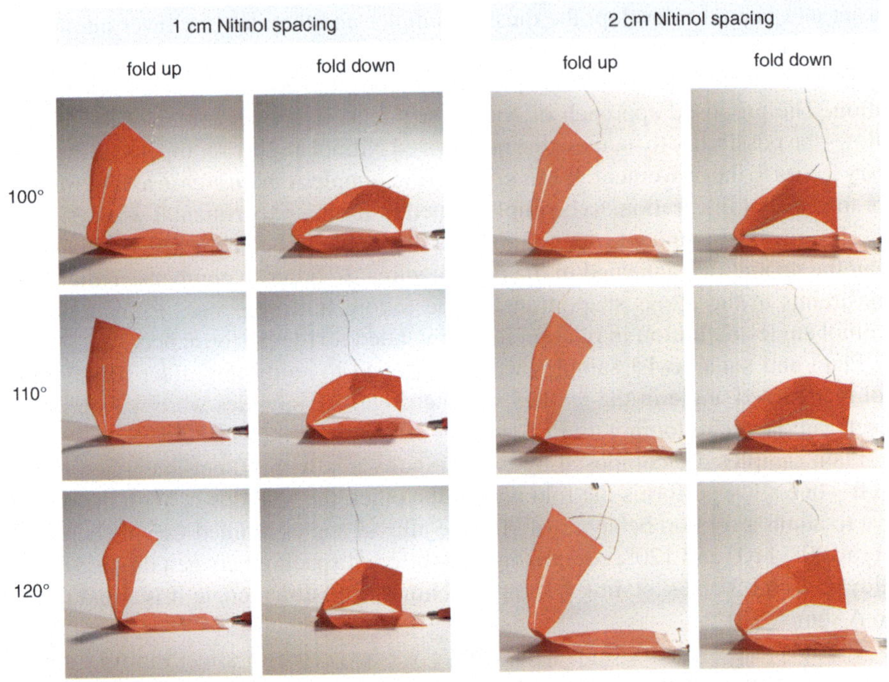

Fig. 6.32 Study samples with different memorized angles and alloys' spacing

up shape in samples with different Nitinol spacing, alloys at 1 cm distance attained angles more similar to 90° than samples with actuators at 2 cm distance, highlighting that the distance between the 2 Nitinol wires with different shapes and activated in different moments, affects the ability to the activated actuator deforms the alloy at M phase. The comparison among the angles defined, samples with Nitinol at 1 cm distance show similar shapes for 110° and 120° samples. In all samples, the fold down behavior presents low accuracy in relation to the Nitinol geometry.

Figure 6.33 presents the shape change results of samples S6 D and E, as well as the images of the previous developed S6 A sample, for a comparison between the activated shapes with and without fold up angle increase. Sample D have improved the angle behavior during fold up in relation to the sample S6 A, but the fold down activation was slightly inferior. Sample E, which presents two fold down actuators per each fold up, had similar performance in the fold down and lower in the fold up in comparison to S6 D.

The behavior of origami samples with equal setup for fold up wires with 110°, 120° and 130° angle, is illustrated in Fig. 6.34. Sample G attained a fold up shape more similar to the predefined, it is possible to observe a slightly more closed effect of the tabs and square area, comparing to samples D and F. The fold down was relatively similar in all samples, although the morphology is not as flat as the predefined.

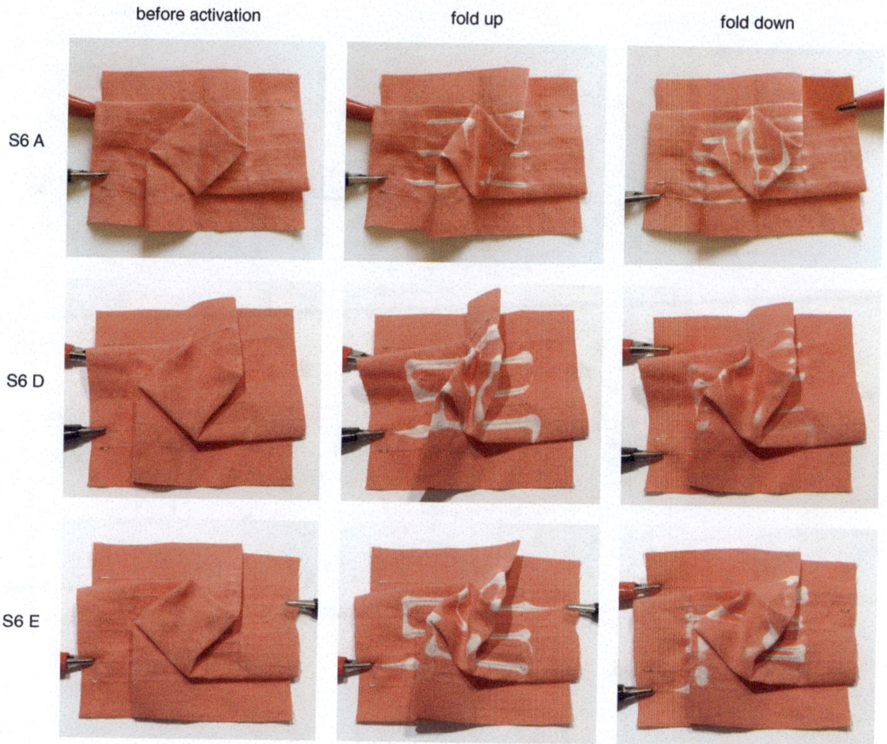

Fig. 6.33 Samples behavior comparison: S6A, D and E

In this study, the improvement between the fold up shape defined and the textile behavior, was attained within the increase of Nitinol wires' angle. However, the fold down was compromised and the actuators performed slightly inferior shape accuracy. The sample with two fold down actuators for each fold up Nitinol did not promote a better performance. Taking into consideration that in this model setup, a decrease of the alloys' angle creates a cross point which would affect the Nitinol integration in the weft direction, the change of the fold down actuator angle was not addressed.

In conclusion, the change of Nitinol angles in relation to the textile morphology, affects the behavior of each actuator group and can be selected in regards to the final objectives. In the present research, the compromise between a greater accuracy of the fold up in relation to the fold down behavior also took into consideration the aim to create textile layers variation. When the tabs are orthogonal to the textile sample, a large area with 1 textile layer is attained. If the textile is able to perform the fold down behavior, the decrease of accuracy interferes in the layers distance but the final morphology layers' number is attained, thus higher light transmittance variation is possible if the fold up is more accurate than the fold down.

Previous origami samples researched the textiles shape accuracy within crease patterns of equal dimensions—6 cm square diagonal. The follow experiment studies

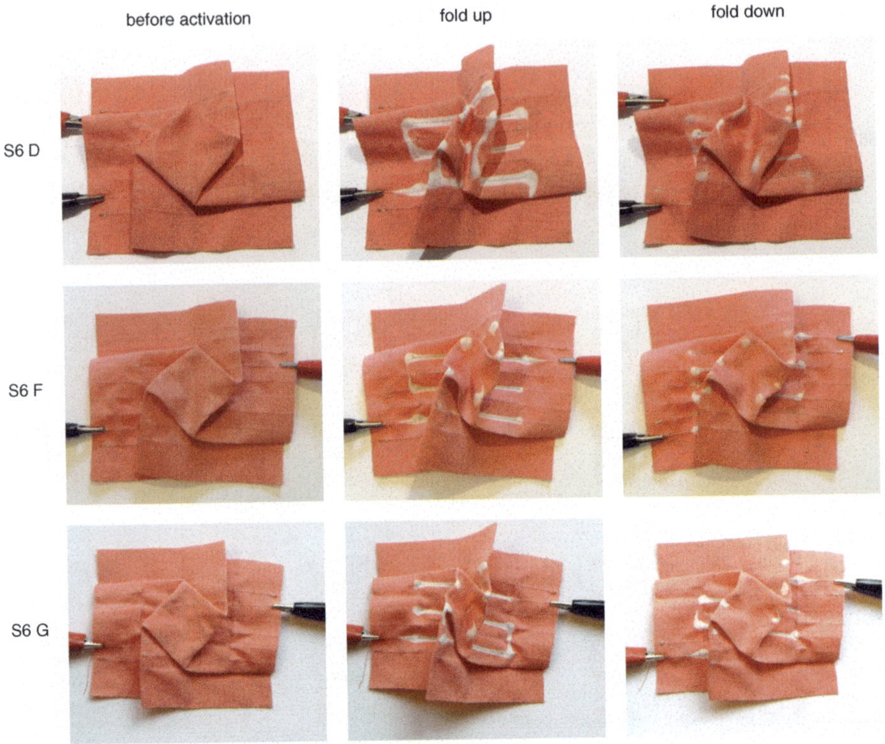

Fig. 6.34 Samples behavior comparison: S6 D, F e G

the shape change development and performance with larger scale shape memory textiles.

Two squaretwist samples were produced following the S6 G process development and Nitinol fold up angle definition—130°. The dimensions selected include a 12 cm square diagonal sample (S12) and a second sample with 18 cm (S18), thus double and triple the previous square diagonal size. Their crease pattern and Nitinol integration setup are depicted in Fig. 6.35. Spacing between Nitinol alloys is 1 cm for S12 and 1.5 cm for S18.

The samples activation was conducted with a Velleman power supply and 0.7 A was applied to each actuator group. The performances were recorded on video and the frames of each activation stage were compared across the 6, 12 and 18 cm square diagonal samples.

In Fig. 6.36 it is possible to observe the samples' scale in comparison to the dimensions of the previous S6 samples. S18 presents a dark color in the laterals due to the loom warp setup with color yarns.

Figure 6.37 presents the samples' images before activation and at fold up and fold down stages. It is possible to observe that the fold up activation shape was more closed in the square area, from smaller to larger samples, while the fold down was

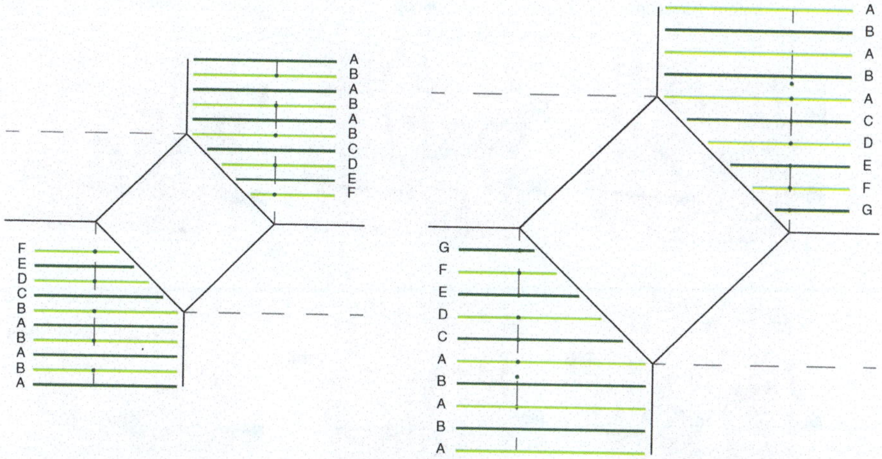

Fig. 6.35 Crease pattern of S12 sample (left) and S18 sample (right)

Fig. 6.36 Samples scale S6G, S12 and S18

more open in S18. Although S12 and S18 present twenty Nitinol alloys each and S6 G has twelve, the wires number increase was not sufficient enough to attain a similar fold up behavior. The size of the samples with the same woven cloth set, are directly related to their weight, thus interfering in the forces required for the SMAs to perform the shape transfer. Besides, as the samples were placed in horizontal position, the fold down behavior is enhanced with heavier textiles. Nitinol angle and integration setup should be optimized, regarding the samples characteristics.

Across the morphologies attained with fold up and fold down activation, it was also possible to observe shape change when the actuator group which was heated up, cools until M temperature and the other group is not immediately activated. This effect was discreet in S6 samples, but in S12 and S18 was evident.

Figure 6.38 presents the shape stages of activation and pause after it, for S18 sample placed in horizontal position (1st row) and vertical position (2nd row). The fold up was activated during 10 s, followed by a pause of 10 s and the fold down and pause after it, with the same time durations.

In horizontal position, the shape attained during the pause after fold up is more closed than the pause after the fold down, meaning that the previous activation shape

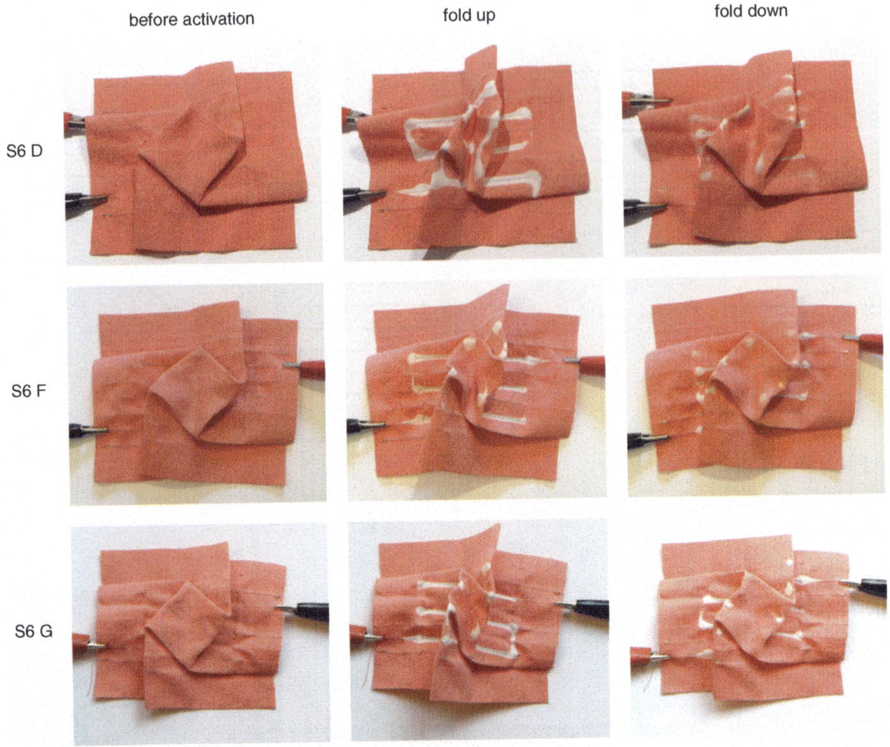

Fig. 6.37 Samples behavior comparison: S6G, S12 and S18

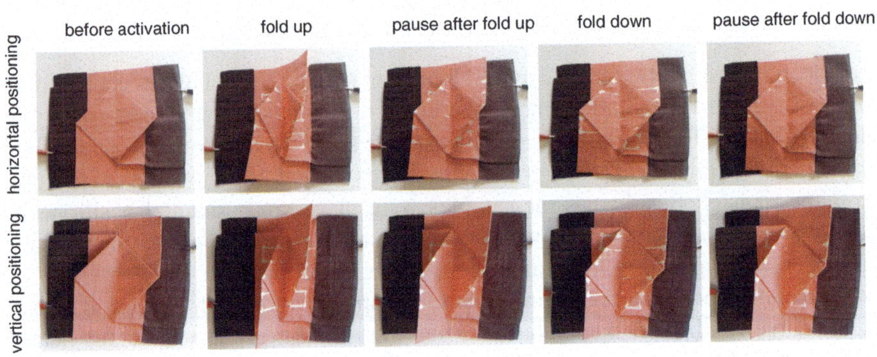

Fig. 6.38 S18 sample behavior in horizontal and vertical positioning

influences it. Comparing the horizontal and vertical position, the fold up is slightly more closed when the sample is placed in vertical position and less open during the fold down, highlighting the influence of the gravity force in the sample behavior.

Samples scale and placement affect the overall shape change behavior at the activation stages as well as after activation. During the SMAs activation pauses, the textile structure assumes a shape in relation to the previous activation stage, the two actuator groups' forces at M phase, their positioning and host structure characteristics. This study attained an understanding of the shape change performance in the selected morphology and scales and highlighted the optimization requirement of the Nitinol geometry and textile integration towards shape memory textiles design.

6.7 Conclusions

In this study, a novel approach was conducted in regards to: Nitinol wire shapes being defined accordingly to a textile geometric morphology; integration of Nitinol segments with different shapes in the same woven textile structure; and interaction of the textile SME with light transmittance.

Design and development of shape memory textiles encompass a systematic process. The material research conducted outlines a process to develop shape memory textiles with the ability to perform two shapes change behavior similar to predefined morphologies.

The workflow setup consists of: defining the origami morphology and shape change; selecting the Nitinol wire and defining geometries, depending on their integration in the woven structure; drawing and production of the dies; optimizing the annealing parameters following the Nitinol heat treatment and finishing processes; and integrating Nitinol wires in the weaving process.

The proposed bias mechanism demonstrates potential to perform shape change for two different morphologies without application of external force, considering a resistive heating activation method. However, the relation of SME constrains between actuator groups and textile substrate is more complex than in textiles that just perform shape change behavior from one temporary to a memorized shape. Within the origami samples developed, the results attained suggest that a compromise on the definition of one of the memorized shapes appear to be necessary.

Additionally, the results have demonstrated the influence of the samples dimensions, weight and positioning in the overall shape change behavior. According to the design requirements and textile characteristics, the Nitinol wires' angle and integration in the woven structure based on the main crease axis areas, should be optimized towards greater similarities between predefined and resulting morphologies, as well as to specific application goals.

Textile morphology definition, based on origami techniques to define Nitinol geometries, was found to be successful in the design of shape memory textiles with dynamic light filter behavior. The methodological processes conducted and the results attained through the samples developed and their behavior analysis afford an informed understanding to design shape memory textiles and set a framework aimed to be implemented at future developments in the design research.

References

Bengisu M, Ferrara M (2018) Materials that move: smart materials, intelligent design. Springer International Publishing, Cham

Berzowska J, Coelho M (2005) Kukkia and Vilkas: kinetic electronic garments. In: 9th IEEE international symposium on wearable computers, 18–21 October 2005, Osaka, Japan

Cabral I, Souto AP (2015) Shape memory textiles: screen printing and knife coating processes to acquire functional performance. In: Blaga M (ed) Proceedings of AUTEX 2015 world textile conference, 2015 Bucharest, Romania

Cabral I, Souto AP, Carvalho H, Cunha J (2015) Exploring geometric morphology in shape memory textiles: design of dynamic light filters. Text Res J 85(18):1919–1933. https://doi.org/10.1177/0040517515578328

Coelho M, Zigelbaum J (2011) Shape-changing interfaces. Pers Ubiquitous Comput 15(2):161–173. https://doi.org/10.1007/s00779-010-0311-y

Dyer PE (2010) Dynamic control of active textiles: the integration of nickel-titanium shape memory alloys and the manipulation of woven structures. PhD thesis, University of Brighton

Hesselbach J (2007) Shape memory actuators. In: Janocha H (ed) Adaptronics and smart structures: basics, materials, design and applications. Springer, Berlin

Lee S (2005) Fashioning the future: tomorrow's wardrobe. Thames & Hudson, London

Liu X, Wang Y, Yang D, Qi M (2008) The effect of ageing treatment on shape-setting and superelasticity of a nitinol stent. Mater Charact 59:402–406. https://doi.org/10.1016/j.matchar.2007.02.007

Rao A, Srinivasa AR, Reddy JN (2015) Design of shape memory alloy (SMA) actuators. Springer International Publishing, Cham

Rottiers W, Van Den Broeck L, Peeters C, Arras P (2011) Shape memory materials and their applications. Korolev's Readings, 2011 Samara, Russia. Samara State Aerospace University

Seymour S (2008) Fashionable technology: the intersection of design, fashion, science, and technology. Springer, Wien

Seymour S (2010) Functional aesthetics: visions in fashionable technology. Springer, Wien

Shabalovskaya S, Anderegg J, Van Humbeeck J (2008) Critical overview of Nitinol surfaces and their modifications for medical applications. Acta Biomater 4:447–467. https://doi.org/10.1016/j.actbio.2008.01.013

Vasile S, Grabowska KE, Ciesielska IL, Githaiga J (2010) Analysis of hybrid woven fabrics with shape memory alloys wires embedded. Fibres Text East Eur 18(1):64–69

Wan T, Stylios GK (2007) Shape memory training for smart fabrics. Trans Inst Meas Control 29:321–336. https://doi.org/10.1177/0142331207069479

Part III
Dynamic Light Filters

Chapter 7
Design Research

Smart textiles are able to interact with the environment and perform dynamic changes over time. Introducing new reactive and adaptive qualities to conventional textiles' functions and expressions, smart textiles open up innovative potential for the design of responsive environments, as they are able of sensing and responding to stimuli and they are an active element in shaping the environment.

The phenomenon approached in this research focuses on the design possibilities of smart textiles to filter light that passes through them, creating responsive lighting scenarios according to textiles physical and dynamic qualities. This interaction was explored through color and shape thermo-responsive textiles, towards changing light transmittance without acting upon the light source: Dynamic Light Filters.

Through a materials research perspective, TC and SMAs behavior was studied and integration processes in textile substrates were developed. Interaction of textile dynamic color with light has demonstrated the ability to transform light transmittance from similar to different luminosities and tones as well as the inverse, in respect to the textiles chromic behavior developed with defined color ratios (Chap. 3). Textile shape change behavior based on origami tessellation morphologies displayed light changes through the variation of the textile layers number, also attaining a change of the light transmittance pattern in the textile surface with geometric areas of different luminous intensities (Chap. 6).

Conductive threads and pigments integrated in textiles were studied for electrical activation of color change (Chap. 4). The experimental results have demonstrated that the electrical current value and the materials' electrical and expressive properties affect heat transfer and textile thermal expansion, highlighting the importance of these variables in how the textile changes color. Conductive materials can present different visual properties when observed in light transmittance and through folding processes their performance can be compromised. The design criteria that in parallel to the electrical and expressive properties, guided the conductive thread selection for the follow research. Nitinol alloys are electrically conductive and shape change aimed of being studied through their resistive heating properties.

I. D. Cabral et al., *Dynamic Light Filters*, Springer Series in Design and Innovation 3, https://doi.org/10.1007/978-3-030-39529-2_7

The results and the findings obtained in the previous chapters created a base framework for the development of smart textiles, which was implemented in a practice-based design research, discussed in this chapter. The objective was to explore dynamic qualities of thermo-responsive textile behavior based on selected variables that affect resistive heating and textile thermal expansion, as well as to discuss the dynamic relationship between textile color, shape and light in the expressions of change.

7.1 Textile Dynamic Behavior

A central subject in smart textiles design involves the changes between states and how the textile performs over time. Designing with temporal forms extrapolates the physical medium definition to embrace the dynamic qualities of the textile behavior, as they play crucial roles in the expressions and functionality of the design outcomes (Worbin 2010; Vallgårda 2014).

Being an emergent dimension in textile design, the introduction of time and movement as inherent dynamic variables presents opportunities and challenges for materiality and methodology, entailing new perspectives in respect to how the textile changes and what it expresses, as well as how they are designed. Therefore, smart textiles require a new understanding of the variables and methods that the designer can work with and develop to explore textile dynamic and expressive possibilities (Redström 2010; Mossé 2016). For example, dynamic qualities are found more comprehensively in disciplines and natural phenomena where temporal dimensions are inherent. Some researchers evolved their methods by seeking and drawing relationships to dynamic phenomena such as in dance (Loke et al. 2015) and music (Jansen 2013), as a way of thinking and exploring time and movement-based concepts in smart and interactive surfaces.

To study textile dynamic and expressive qualities, design possibilities are investigated through the variables that have an active role in the temporal forms and how they are articulated. In this research, the focus lied on temperature as a dynamic design variable to explore qualities and expressions of change in color, shape and light performances.

How temperature varies is a critical consideration with the materials selected. They present gradual transformations during a temperature interval both on heating and on cooling. When researching thermo-responsive textiles electrically activated, temperature involves three perspectives: temperature of the conductive materials, temperature of the textile and room temperature. For each of these perspectives or dimensions, diverse variables can account for thermal variation, composing a set of relationships that can influence textile behavior qualities and thus how light is transformed.

In working with two thermo-responsive materials in the same textile structure, the designer has to consider thermal variation for both behaviors, according to the materials' activation temperature selected. Possibilities to create simultaneous or independent changes can be explored to extend textile and lighting dynamic qualities.

The practice-based design research presented in this chapter proposes to study color and shape behavior of thermo-responsive textiles, with the objective to discuss and demonstrate dynamic qualities of change, based on: the relationship between design variables that the designer can consider and act upon to influence how textile temperature varies; the interaction with light that passes through the textile and how it can transform and be transformed through light luminosities and tones; how textile and light behavior can unveil expressions with different intensity levels.

The research program included two experimental studies and the development of three research prototypes. The initial experimental work has explored the influence of selected design variables in the dynamic qualities of textile behavior and their interaction with light. The first study focused on textile color and shape and the second study included interaction of textile and light. The research prototypes were developed with the objective to explore the relationship between the variables as a system in the expressions of color, shape and light change and to discuss new possibilities and challenges when designing dynamic behavior of thermo-responsive textiles.

Temperature as a dynamic design variable encompasses several relationships between the electrical activation definition, physical context and textile characteristics. The following sections discuss the experimental studies conducted, aiming at analyzing how a set of selected variables influences temperature variation and their relationship with dynamic qualities of textile color and shape behavior.

Considering the diverse interdependencies that can influence temperature variation, each experiment focused on one independent variable at a time and defined other variables to maintain a constant. Design criteria for variables selection was framed through initial questions raised on textile behavior. The definition of each test parameters is presented in a graphic setup and room temperature was approximately 20 °C in all tests, except when specified otherwise.

Color change studies comprise of six experiments and were developed with two electroconductive samples screen printed with TC pigments with equal substrate characteristics, only varying dimension and morphology (sample X and Y). Shape change was studied with one sample (Z) and comprises of four experiments. Sample X and Y development is described in Appendix B and sample Z in Chap. 6.6 (sample S12). Activation temperature of the TC pigment used is 27 °C and Nitinol alloys is As 30 °C and Af 45 °C.

To document and analyze temperature variation and textile behavior, each experiment was recorded via thermal images and video. For a visual communication of the changes over time, thermal images and video frames were organized in a framework, according to a defined timeline.

7.1.1 Color Change

Experiment A aimed to study how textile chromic behavior can be influenced by the definition of an electrical current value for thermal activation. Two tests were conducted. A1 analyzed color change transitions on heating and on cooling for a defined time period. A2 investigated how slow and how fast the textile can change

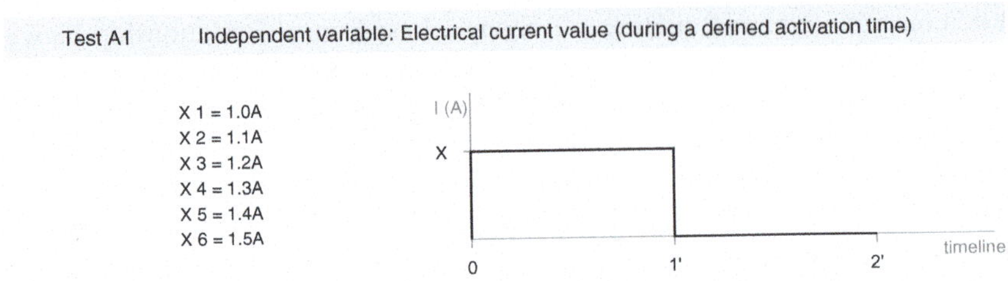

Fig. 7.1 Test A1 setup

from colorized state to complete colorless areas and the inverse, from total colorless to full color return.

For test A1 setup (Fig. 7.1), the duration of power supply was 1 min, after which color return was also analyzed during the same time period. The electrical current values tested ranged between 1.0 and 1.5 A with 0.1 A interval each.

Textile changes over time are illustrated in Fig. 7.2 framework, displaying the visual record of thermal variation (IR images) and chromic behavior (video record frames) at each 15 s of the timeline, for each electrical current level. When the power supply is switched on, the conductive threads' pattern is revealed through the transition from textile solid color to parallel colorless lines. This initial color change stage is perceived in the first 15″ of activation with all electrical current values tested.

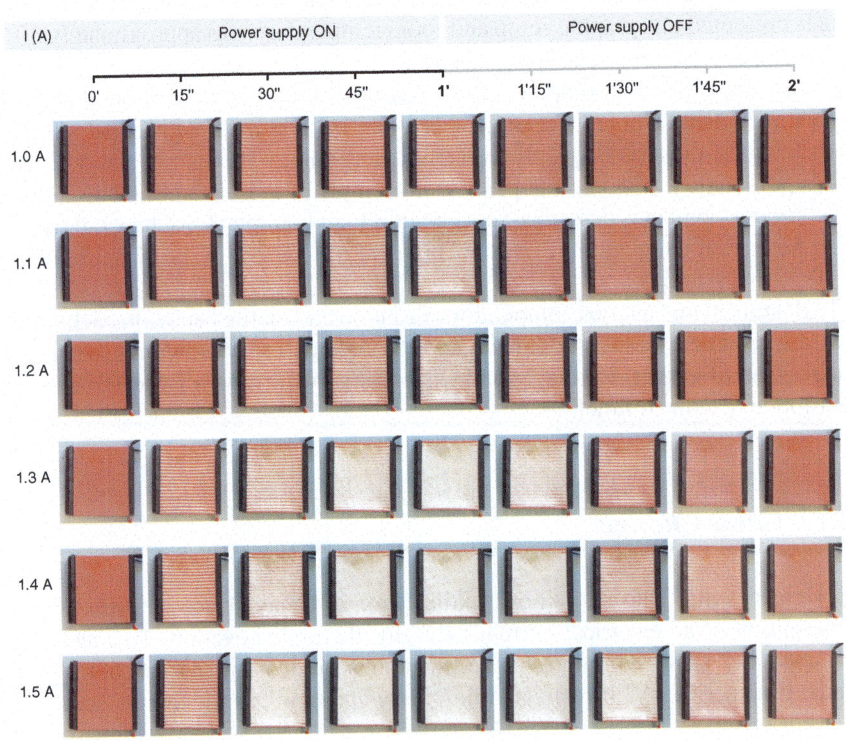

Fig. 7.2 Test A1 results

However, the stripes' thicknesses are already distinct in this phase. With 1.0 A, a thin striped pattern is observed, while with 1.5 A the stripes appear to be as thick as the textile areas still in colorized state.

During 1′ of activation, thermal expansion in the textile areas in between the conductive threads increased in a noticeable faster pace with higher currents. Activated with 1.3 A or more, the sample attained an overall colorless effect, whereas with inferior power, the stripped pattern was still perceived.

When power was interrupted, textile temperatures were different for each test, influencing how the textile changed to colorized state. Besides affecting the time required to cool down, it was also observed that color return exhibited a sharper edge in the linear pattern in tests with lower electrical current values. With higher values the colorization trajectory showed less relevance of the conductive lines. The stripes were still perceived, although they displayed a blurred effect with the overall shades of color return.

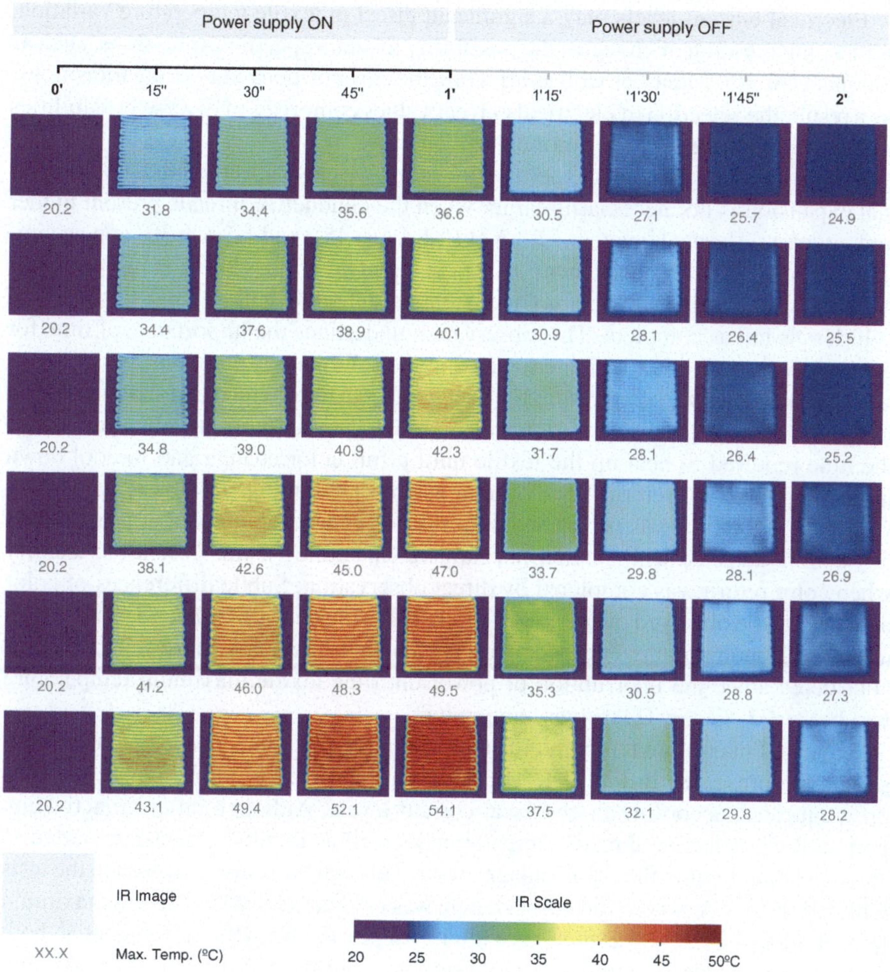

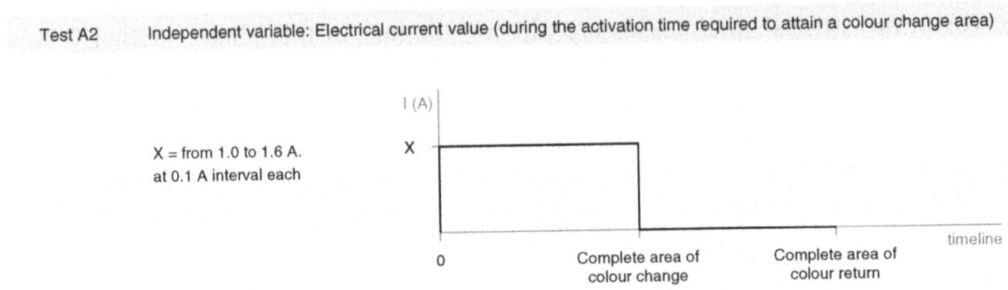

Fig. 7.3 Test A2 setup

Considering the direct observations of the overall performances, perception of the first stage of change from solid color to stripe pattern is evident. During thermal expansion, the gradual changes of color through the colorless lines thickness increase are slower and not as obvious.

Electrical current levels play a significant effect in textile temperature variation, influencing duration of changes, the possibility to attain specific expressions, such as complete colorless areas, as well as affecting the chromic behavior during transitions. As a result, the selection of electrical current values comprises of diverse possibilities to explore and design textile color change qualities.

Furthermore, through this experimental data it was also observed that textile thermal expansion is not necessarily larger when the conductive threads present higher temperatures. Thermal images of 1.3 A at 1′; 1.4 A at 45″; and 1.5 A at 30″ appear similar. Although, for the stages mentioned, a complete colorless effect is just observed with 1.3 A, presenting a lower maximum temperature than with 1.4 and 1.5 A, where colorless is not as complete. This observation underlines the importance of time for color change qualities, a variable focused in the following experiment.

With the objective to explore textile paces between two defined expressions, experiment A2 studied how the activation with constant electrical current values influences the time required to heat up the textile until a full color change and to cool down until complete color return. Test A2 setup is presented in Fig. 7.3.

Textile activation was performed until maximum temperatures of approximately 50 °C were detected with the thermal camera. On cooling it was difficult to identify when color return was completed by direct observation. Subtle differences of color nuances can be observed during several minutes, which can differ between activation tests in the same sample. For a comparative analysis, the color return trajectory was considered since the interruption of power until the textile maximum temperature was below 27 °C, the TC trigger temperature.

Time and maximum temperatures attained in the predefined stages of color transitions are presented in Fig. 7.4. The temperature increase during power presents different rates, according to electrical current levels. Although, after an activation period, the conductive threads temperature as well as thermal expansion started to stabilize, maintaining the color change effect. This behavior was analyzed in the tests with 1.0 and 1.1 A, where full color change was not attained. With 1.0 A, a maximum temperature of 39.3 °C was measured after 5′ of power and in the subsequent 5′, both temperature and textile expression were similar over time. With 1.1 A, the maximum temperature increased only 1 °C from 5′ to 10′.

	Heating: Time	Heating: Max. Temp. (°C)	Cooling: Time	Cooling: Max. Temp. (°C)
1.0 A	(10')	39.0°C	2'	25.6°C
1.1 A	(10')	43.7°C	2'	26.8°C
1.2 A	2'15"	46.7°C	1'45"	26.9°C
1.3 A	1'30"	48.4°C	1'30"	26.8°C
1.4 A	1'15"	50.1°C	1'15"	26.9°C
1.5 A	45"	53.1°C	1'30"	26.6°C
1.6 A	30"	53.1°C	1.40"	25.9°C

Fig. 7.4 Test A2 results

This analysis highlights the possibility to design color change behavior, where colorless expression can be maintained during long periods without overheating the textiles. For this performance, a relation between optimum current level and temperature in the conductive threads that attains constant color change areas requires experimentation.

Observing the results attained with 1.2 A, the textile became decolorized at $2'15''$, with only 3 °C maximum temperature higher than with 1.1 A. Whereas maximum temperatures were similar, an analysis of the thermal images of the stages mentioned showed that with 1.1 A few areas attained the maximum temperature, observed through orange and red color of the IR chromatic scale, while the thermal image with 1.2 A presented a more homogeneous distribution of the high temperatures in the textile. Nevertheless, mean temperatures in the textile areas between the conductive threads were not very distinct, reflecting the impact that small temperature differences can have on color change expressions and the importance of understanding these relationships to design thermo-responsive behavior.

Color return after switching off power supply at the complete colorless state is also influenced by the relation between electrical current level and time of activation. Comparing the tests from 1.2 to 1.6 A, time on heating until full colorless decreases and maximum temperatures increases, whereas the time of color return transition decreases from 1.2 to 1.4 A tests and increases from 1.4 to 1.6 A. The cooling effect is not only associated with the maximum temperatures attained but with the overall textile temperatures that affect how the textile returns to colorized state.

Higher electrical current values allow for designing faster color change behavior, while color return pace is also affected by the previous period of power supplied.

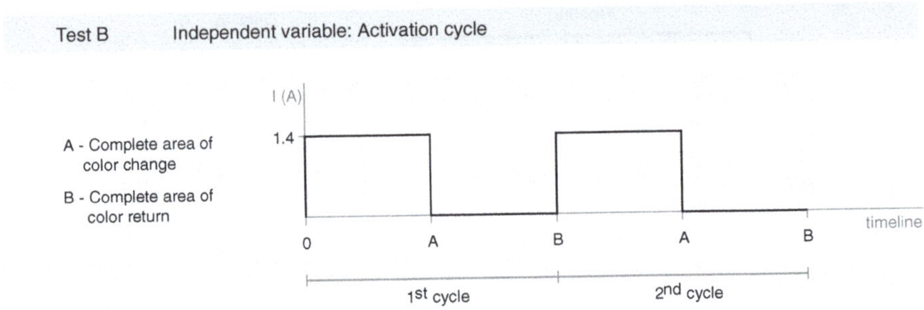

Fig. 7.5 Test B setup

Considering the setup defined for this study, 2′15″ with 1.2 A was the slower color change on heating and 30″ with 1.6 A was the faster. Whereas the graphic shows data until 1.6 A, higher values were also tested, attaining temperatures above 50 °C

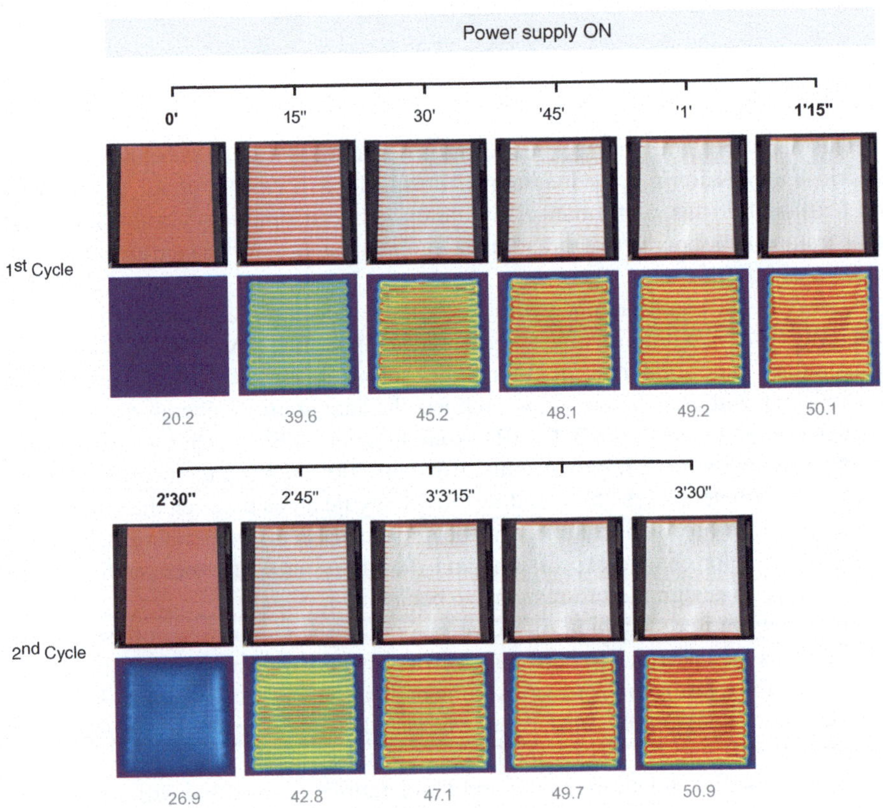

Fig. 7.6 Test B results

in the conductive threads before complete colorless area was observed, which set the limit for this test analysis.

Color change activation may comprise of different sequences of textile heating and cooling down. Test B aimed to compare textile temperature variation and resulted color change effects of two consecutive cycles of full color change and full color return. The electrical current value defined for experimental B setup was 1.4 A (Fig. 7.5).

The images framework presented in Fig. 7.6 enable a visual comparison of the results attained in the two consecutive cycles of activation. The sample was in colorized state in the beginning of each cycle, but their temperature differences influenced the textile behavior. The textile was warmer at the starting of the 2nd cycle than in the 1st cycle and attained a decrease of color change duration until colorless of approximately 15″ and an increase in the time of color return also approximately 15″.

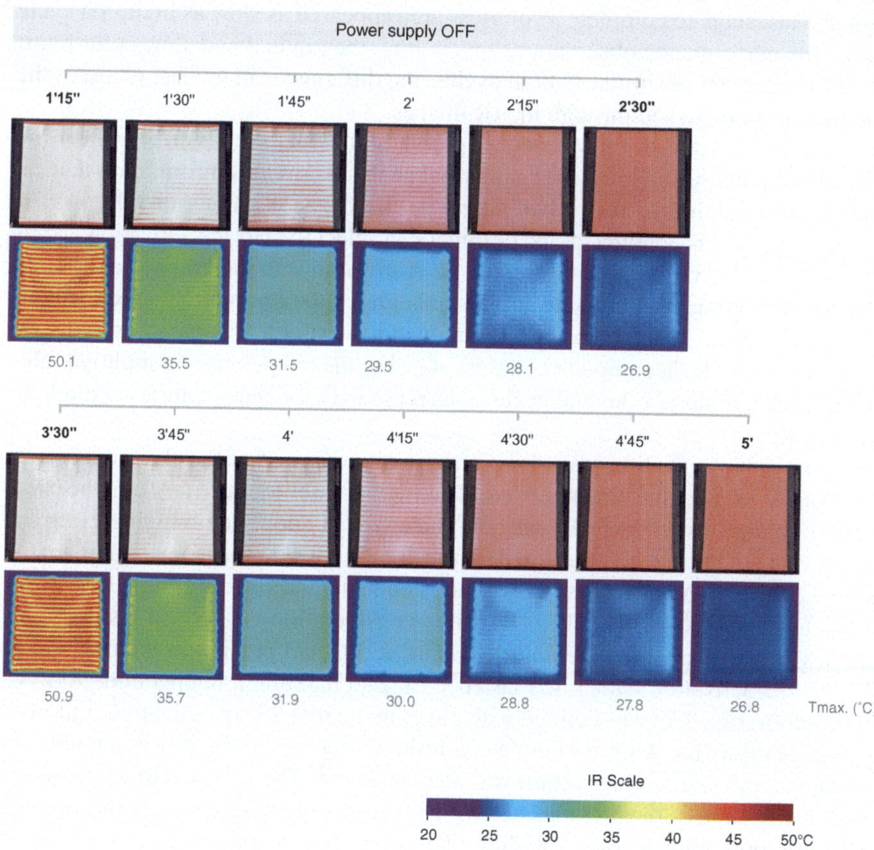

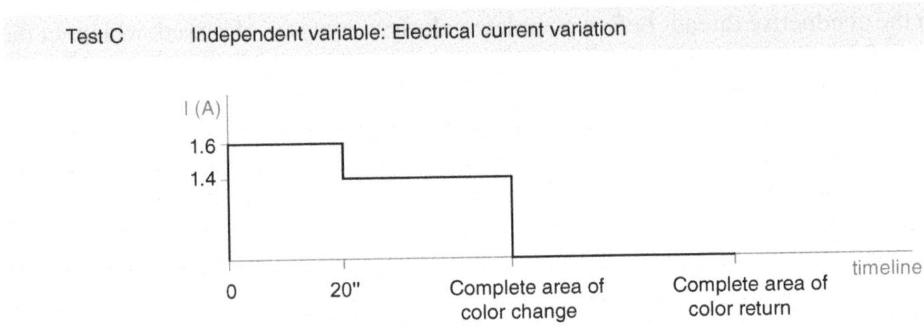

Fig. 7.7 Test C setup

Visually, the color behavior in the 2nd cycle was particularly affected in the beginning, where the stripe pattern rapidly showed the increase of the line thickness, while the last stage to complete a colorless area appeared as slow as in the 1st cycle. During the performance observations on cooling, the gradual and slow color return was also perceived as similar in both cycles, the differences in the last phase of this transition were observed through the IR images.

To explore and design color change expressions, the focus relies on how the visual changes are perceived. However, the thermal analysis provides information that can be used as an additional tool when taking design decisions. For example, if a 3rd consecutive cycle is conducted, the designer can select to start the activation at 5′ or later, so the paces of color change could be more similar to the 2nd cycle, or to the 1st cycle if the time without activation is enough for the sample to cool down until room temperature. If activation is programmed to start at 4′45″, where visually the textile appears to be at complete colorless state but the temperature is slightly higher than in the beginning of the 2nd cycle, a decrease in color change duration might be expected.

Color change activation may also comprise of a variation of the electrical current levels supplied. With the objective to study design possibilities to transform the paces of color change, experiment C analyzed the effect of supplying a higher electrical current value in the first stage of textile activation, followed by a decrease until textile thermal expansion in the areas between the conductive threads is complete.

The test definition and analysis took into consideration the A2 test results. Activation with 1.6 A attained the colorless state at 30″ and maximum temperature of 53.1 °C; the activation with 1.4 A lasted 1′15″ and maximum temperature 50.1 °C. Test C setup (Fig. 7.7) was defined with 1.6 A in the first 20″ of activation, followed by a decrease to 1.4 A until complete colorless after which the power supply was interrupted and textile color return was also analyzed. The aim was to decrease the color change time of the previous test with 1.4 A, maintaining the color return time.

Results presented in Fig. 7.8 show the color change transition until colorless attained in 1′ and color return until 2′30″. Comparing the results attained with A2 test 1.4 A, there was a decrease of 15″ on heating with a slight increase of the maximum temperature and a 15″ increase on cooling. The results can also be compared with the images of A1 test where with 1.4 A in 1′, small color change differences can be observed in some areas of the textile margins that are still colored.

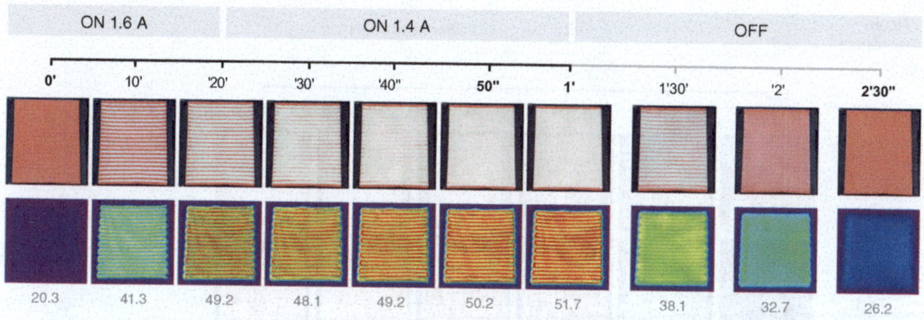

Fig. 7.8 Test C results

If the design parameters focus on the overall color change duration, activation just with 1.6 A can be selected, as it was significantly faster on heating and presented approximate duration on cooling as well as maximum temperatures.

The potential highlighted by using electrical current variation relied on the possibility of having faster transitions in particular moments of the activation, thus enabling to design different rhythms of color change behavior from on state to another. In this test, expressions of change had a faster pace within the stripe pattern, emphasizing the beginning of the color change transition. If the activation is defined with low electrical current levels in the first phase and higher levels after, slower changes will be created when the performance begins, increasing the rhythm of change that completes the textile colorless effect.

Room temperature is a variable of the physical context where color change occurs. Experiment D studied the effect of two different room temperatures in the textile chromic behavior, with the same current supplied. D setup was defined with 1.4 A activation, to be tested at 16 and 20 °C room temperatures (Fig. 7.9).

As analyzed in the Fig. 7.10 framework, room temperature significantly affects changes of textile color. At 20 °C room temperature, the striped pattern of color change starts to blur into colorless areas at 45″ and at 16 °C room temperature, a similar expression is observed at 1′45″. The decrease of 4 °C in the physical context has changed the time required for the textile attains full color change to be twice as much.

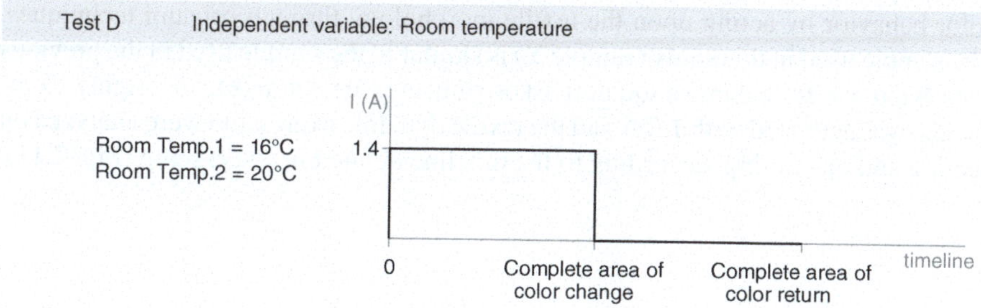

Fig. 7.9 Test D setup

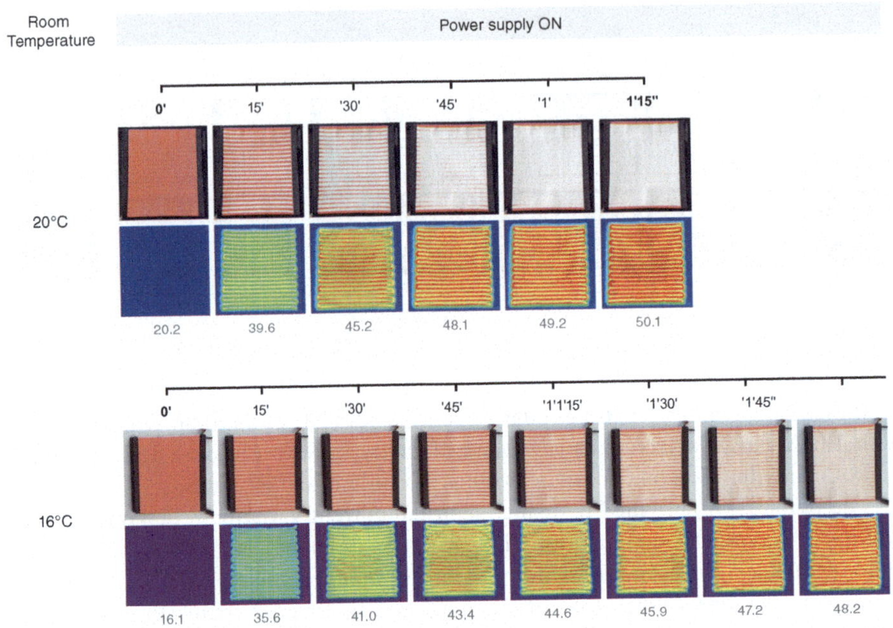

Fig. 7.10 Test D results

The effect on the textile cooling down to temperatures below 27 °C was not as distinct in duration. As previously discussed, the duration of power supplied during activation plays an important role in the color return changes. However, other qualities of color return behavior were affected. At the lower room temperature, color return in the textile areas in between the conductive threads was perceived very fast after the power supply was switched off and the lines' edges appeared more sharp, while at the higher room temperature the colorization changes had a blurred expression.

These results highlight that even in interior environments, where room temperature might present small variations, it can exhibit a perceived effect in the dynamic qualities of the textile behavior. Considerations should be taken when defining and programming the activation sequences, in respect to the textile physical context.

In experiment E the objective was to explore design possibilities of dynamic color behavior by acting upon the textile morphology through origami techniques. The sample used in this study (sample Y) is similar to the sample used in the previous tests, but it was folded into a squaretwist origami crease pattern (Fig. 6.2 right). Color change was activated with 1.2 A and the textile dynamic expression were analyzed on heating and on cooling, according to the timeline defined in test E setup (Fig. 7.11).

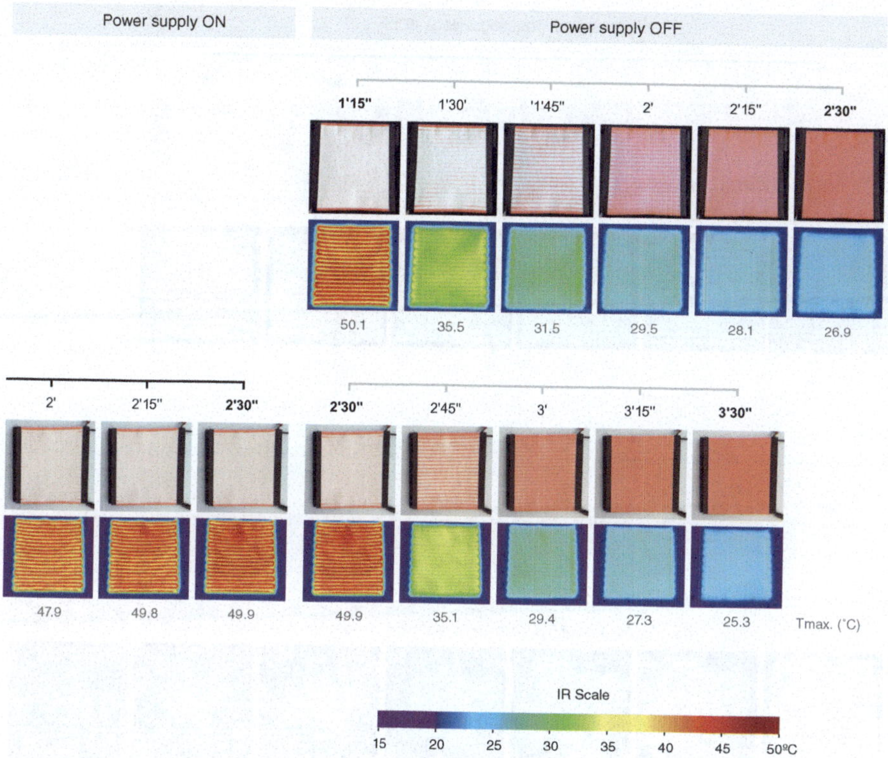

The conductive threads integrated in the sample substrate are parallel. However, the squaretwist folding of the textile has created areas with different pile-up layers and a square area that was twisted, changing the direction of the conductive threads in the textile upper layer, which become perpendicular to the threads in the lower layers. As a result, textile morphology has influenced thermal variation and color change expressions (Fig. 7.12).

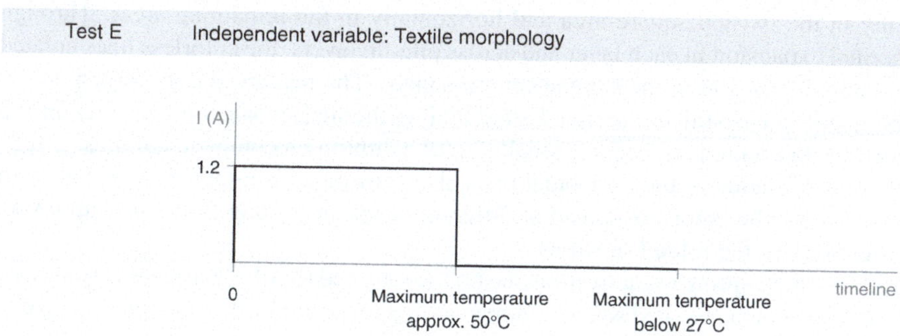

Fig. 7.11 Test E setup

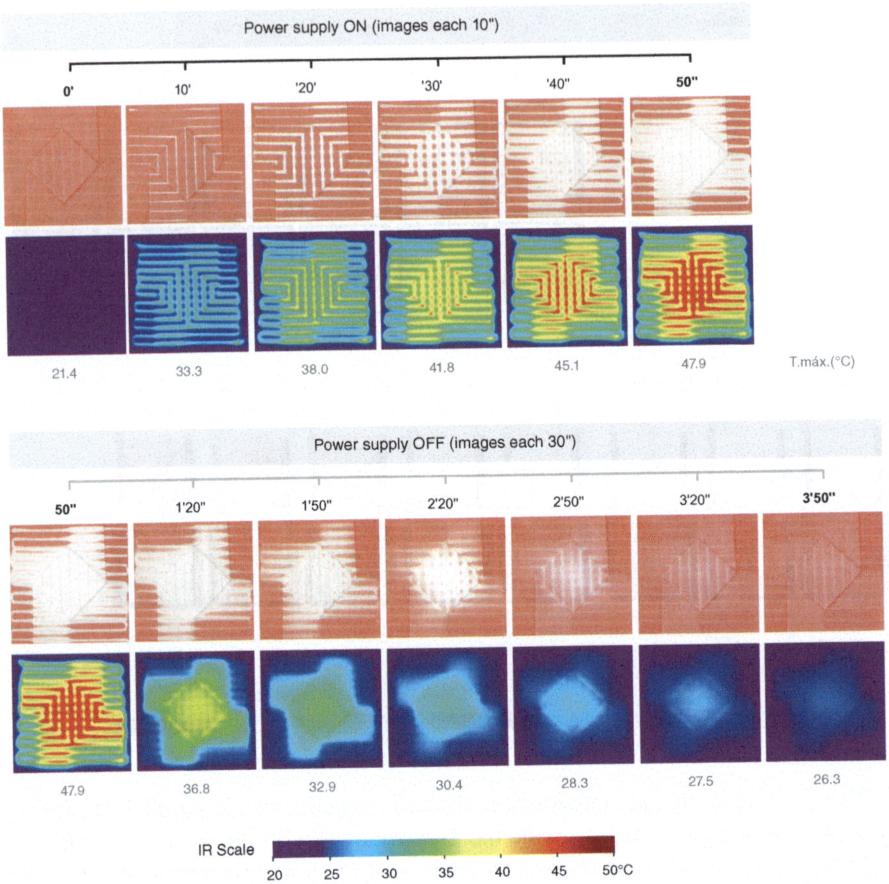

Fig. 7.12 Test E results

In the first stage of the activation, thin lines of color change are revealed vertically in the twisted square area and horizontally in the remaining areas. Through thermal expansion in each layer and across pile-up layers, the colorless lines created a spotty expression in the sample central square. The perceived dots correspond to the textile areas still in colorized state and as thermal expansion increases in the dots' background, they become smaller until exhibiting a complete colorless effect. Thermal variation in the horizontal lines also showed a heterogeneous change with areas where the colorless spread at different paces, according to the pile-up layers differences of the folded substrate.

After interrupting the activation, the textile returned to colorized state very slowly, particularly in the twisted square, corresponding to the area with more pile-up layers.

During the 3′ required to cool down the sample until temperatures below 27 °C, changes between colorless and shades of color appearing presenting a blurred effect and discreet dots in a small phase of the color transition.

This study demonstrates the potential of exploring dynamic qualities of color change through textile origami morphologies. Diverse origami base folds, folded pattern compositions, techniques, scales, etc. can be applied to transform the substrate, changing the geometry of the integrated heating elements, thus enabling the creation of color change patterns with different dynamic expressions.

7.1.2 *Shape Change*

Similarly to the tests conducted in experiment A for color change, experiment F studied how shape change behavior is influenced by the electrical current levels. Sample Z activation from deformed flat shape to fold up was conducted until a maximum temperature of approximately 45 °C (Nitinol Af). After power supply was cessed, textile cool down was analyzed until the textile maximum temperature was below 30 °C (Nitinol Mf), (Fig. 7.13). The electrical current values tested ranged between 0.6 and 0.9 A with 0.1 A interval each.

The sample used to study shape change behavior was screen printed with a TC pigment that has a lower activation temperature than the Nitinol alloys used. As a result, the activation of shape change also changed the textile color, revealing the SMAs integration pattern. At the angle that the tests were recorded, the colorless lines were mostly perceived when the textile folds up, adding a new variable in the perception of the shape change behavior.

The shape change between two states can be designed with different paces according to the electrical current values defined for activation. The study on this relationship is depicted in Fig. 7.14. Through a direct or video record analysis, the slow movement observed with 0.6 A appeared discreet and after a first activation phase, it was not clear if the textile shape was still or slowly changing. The images framework showed that from 15″ to 50″ the shape has changed slightly.

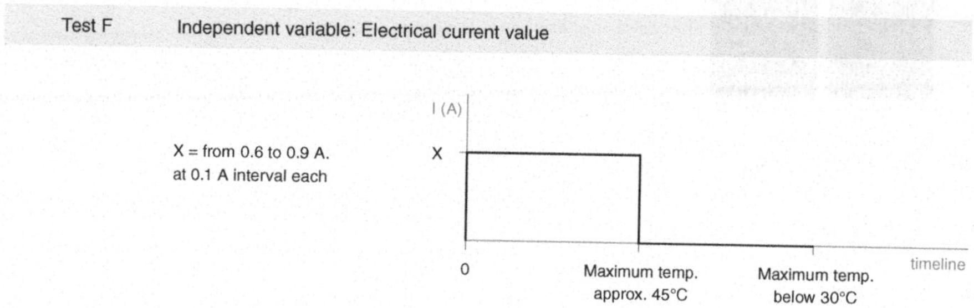

Fig. 7.13 Test F setup

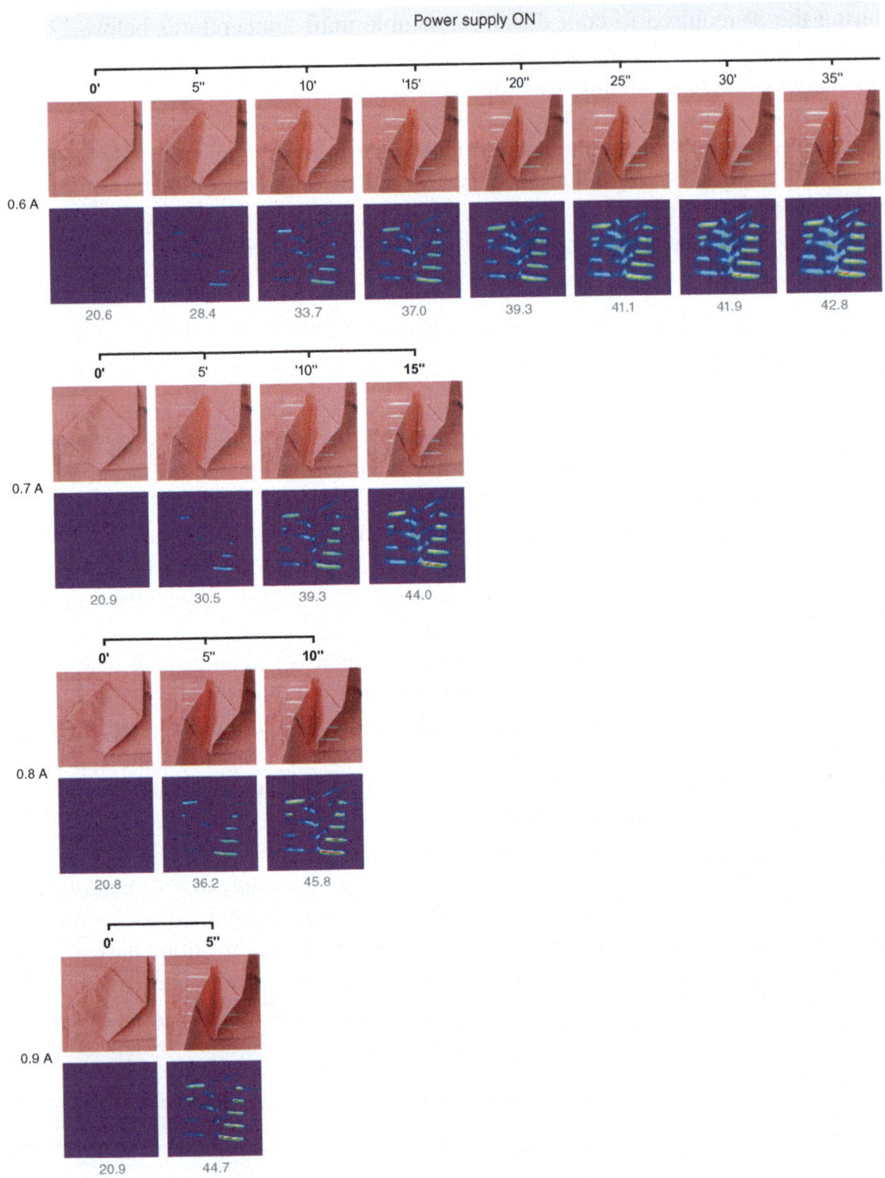

Fig. 7.14 Test F results

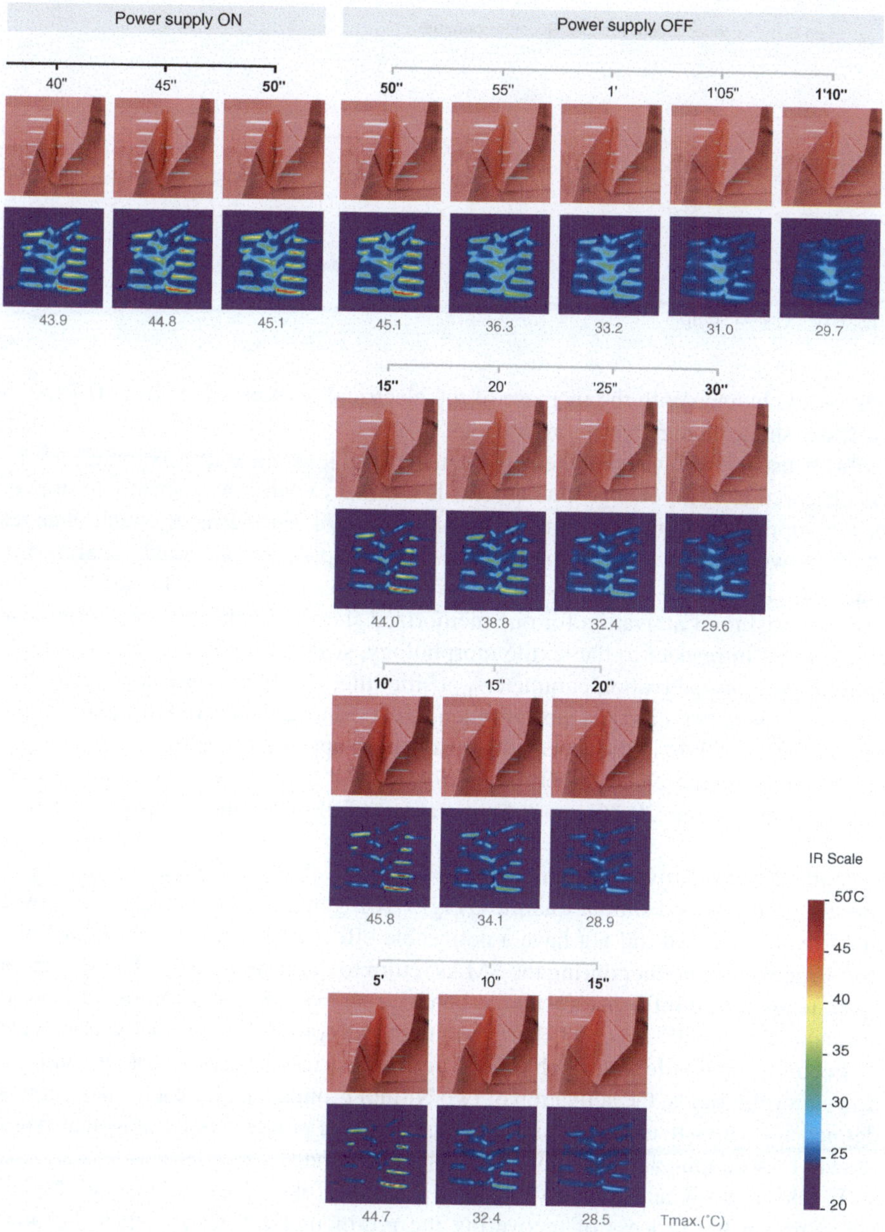

The increase of electrical current levels attained faster shape change, whereas the movements had an organic expression independently of the paces. Thermal variation was faster with 0.7 A than with 0.6 A, but the main perceived behavior appeared to be similar, as can be visualized in the images until 15″. This analysis points out that decreasing the power supplied does not necessarily slow down the pace of the

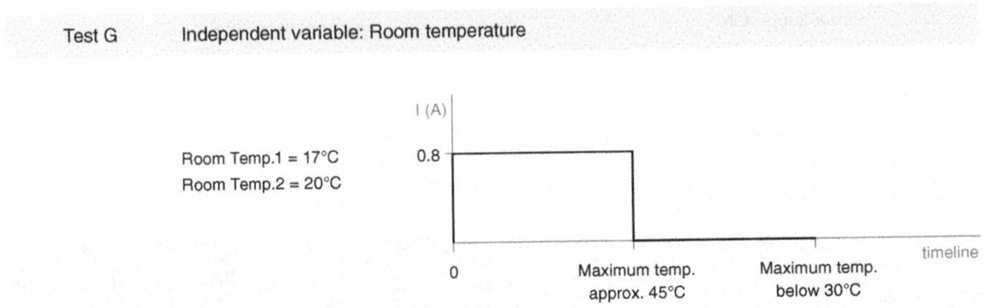

Fig. 7.15 Test G setup

perceived change. With the increase of the electrical current values from 0.7 to 0.9 A, faster shape change was more obvious.

In all tests, power was switched off when the sample attained approximately 45 °C. As observed in color change studies, textile thermal expansion according to time of power supplied has influenced the Nitinol cool down. Furthermore, small changes of shape were observed after interrupting power supply, as previously analyzed in Sect. 6.6. This effect highlights the possibility to design textile behavior that even just comprising of activation for one memorized shape, the changes on cooling can be explored in respect to the textile morphology, scale of the folded motif, sample weight, previous activation parameters, positioning, etc. In the tests conducted, this movement was perceived as slower when the textile was activated during a longer period (0.6 A) while in the other tests the changes appeared similar.

In test G, textile changes between deformed flat morphology and fold up were analyzed at 17 and 20 °C room temperatures, following the setup presented in Fig. 7.15.

With the same activation parameters, textile shape change at the two room temperatures tested attained similar qualities (Fig. 7.16). Temperature increase was slower at 17 than 20 °C but did not have a noticeable effect on the textile temporal forms during activation neither during the SMAs return to deformable state. The studies on room temperature influence on textile behavior showed that this parameter is more critical in the qualities of color changes than shape, considering one activation cycle.

Focusing on textile shape behavior to perform variation between fold up and fold down morphologies, the activation of two Nitinol groups, one for each shape, can be defined according to electrical current values, time of power supply, as well as pause time in between the activation of each memorized shape.

If fold down is activated immediately after the fold up, or the inverse, Nitinol alloys are in the process of recovering the pre-memorized shape when the other Nitinol group is still above the temperature that can be deformed. There is a time duration increase to attain the complete shape change, in regards to the time required for the actuator group previously heated up to cool below 30 °C.

Test H studied the influence of non-activation times between the textile fold up and fold down activation in the qualities of change. The setup of the two tests conducted are depicted in Fig. 7.17, where 0.8 A is supplied for 10 s for fold up and fold down

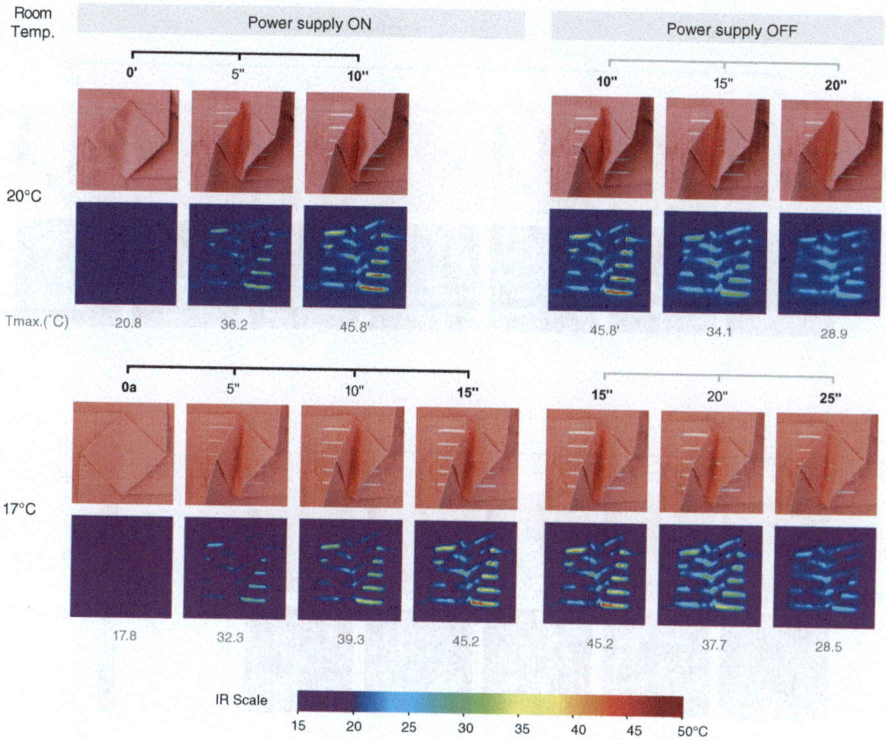

Fig. 7.16 Test G results

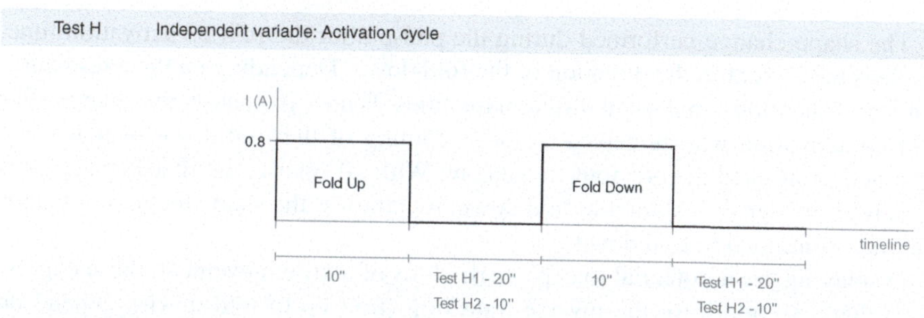

Fig. 7.17 Test H setup

in both tests and the pause time after each activation is 20 s in H1 and 10 s in H2.

Different cooling times create the frame for the changes duration, influencing temperature variation and how the textile changes shape (Fig. 7.18). When fold down was activated, textile maximum temperatures in H1 and H2 tests were below the Mf, meaning that the Nitinol fold up group previously activated was in the deformable state and after 10″ of fold down activation the textile presented similar shapes.

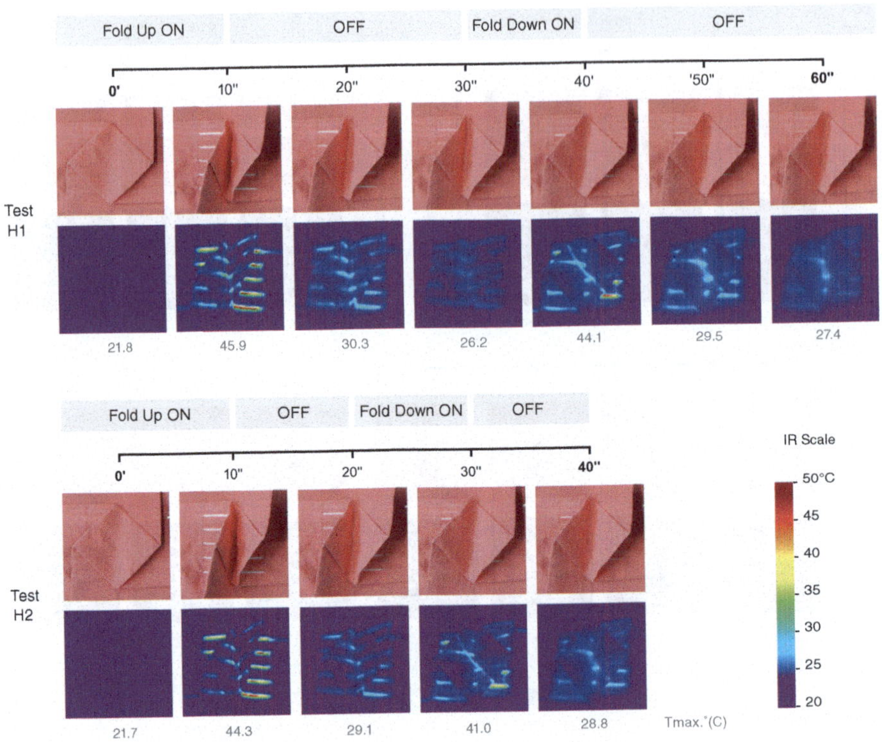

Fig. 7.18 Test H results

The shape change performed during the pause after the fold up activation time, displays movement in the direction of the fold down. Depending on the pause duration, this transition can present different qualities. With a 10″ pause, the change after fold up activation was perceived as the beginning of the fold down, which when activated, continued the previous movement. With 20″ pause, the shape change was perceived in 'steps': it started to fold down, maintained the shape for some seconds and then continued to fold down.

Displaying great potential to explore rhythms of textile movement, these expressions that also apply for the inverse transition (fold up to fold down), depend on diverse parameters. For example in Sect. 6.6, the comparative study of shape change in vertical and horizontal positioning, observed subtle differences between the textiles morphologies after the activation pause. Therefore, the relation between activation and non-activation times requires experimentation with the intended textile structure and placement, whose dynamic qualities aim to be designed.

The sequential activation between two Nitinol groups creates movement that can continuously evolve from one state to another. The objective of experiment I was to study how shape change activation through consecutive cycles may influence textile temperatures and morphologic behavior. Test I setup (Fig. 7.19) was defined with

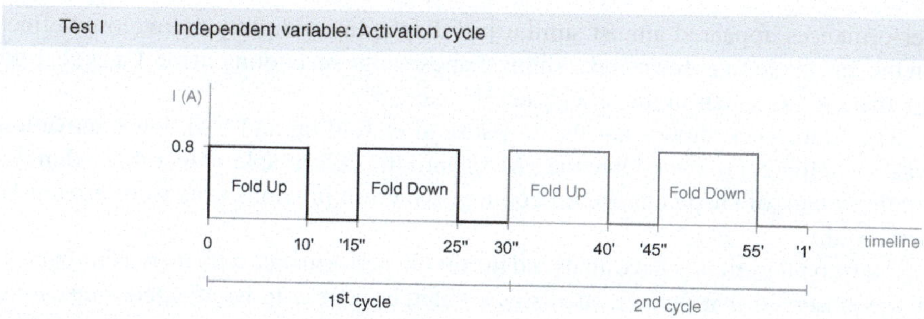

Fig. 7.19 Test I setup

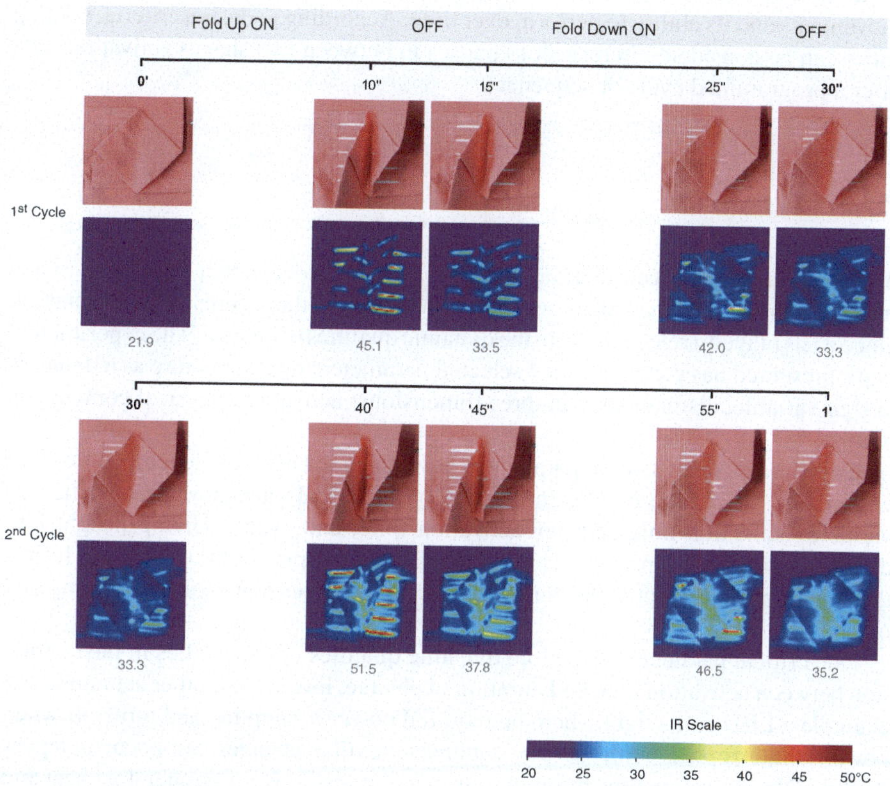

Fig. 7.20 Test I results

two fold up and fold down cycles, supplying 0.8 A for 10 s in each activation stage and 5 s for each pause in between activations.

Activation of two consecutive activation cycles affected thermal variation of the Nitinol alloys and also the textile areas in between them (Fig. 7.20). Temperature differences were mostly emphasized in color change, whereas in shape change the performances appeared almost similar just displaying a slightly more closed effect on the 2nd cycle fold down and a slight shape change on cooling in the 1st cycle fold up, that did not occur in the other paused states.

The framework shows that the 5″ pause after fold up and fold down activation was not sufficient to cool down the SMAs until the deformable state, reflected in the subtle or lack of shape change on cooling, which in previous tests were obviously perceived.

Additional exercises have analyzed the textile performance with increasing cycles. It was observed that the activated shape changes started to be affected, presenting lower abilities to fold up and fold down until the movements were near imperceptible, as both Nitinol groups simultaneously achieved in high temperatures.

Continuous activation sequences significantly affect the qualities of the textile movements and its ability to perform over time. According to design criteria, cooling times can be considered in respect to pauses in between two shapes activations and after a programmed cycle or sequences.

7.1.3 Discussion

The relationship between thermal variation and the independent variables studied have demonstrated to be crucial when designing textile thermo-responsive behavior, since it has played a relevant role in the dynamic qualities of change. The experimental work presented has comprehended selected parameters of temperature as a dynamic design variable, distinguished in three dimensions: activation, physical context and textile.

Experiments on activation parameters have analyzed how textile temperature and expressions of change can be influenced by electrical current value, constant or variable; activation time and non-activation time; and cycles. Giving insight into design possibilities of paces of change, rhythms and expressions of temporal forms, the studies have illustrated the complex interdependences that evolve in heating and on cooling.

One critical parameter to explore dynamic qualities of change lies in the distinction between activation time and duration of change, inversely to other actuators. For example a LED emits light when the required power is supplied and turns off when power is interrupted. With thermo-responsive textiles changes can occur independently of the activation time through a nuance of transitions comprehending heat and the inherent properties of the color and shape change textiles. This quality presents challenges but also opportunities to design smart textile behavior.

Whereas the selection of activation parameters may initially appear to be attributed to a technical decision, their influence in the textile dynamic qualities suggest they are expressive variables of color and shape change. For example, how paces and expressions of color change can vary, when the textile is activated during the same time with different electrical current values.

Thermo-responsive behavior with materials which activation temperature is higher than average room temperature is commonly focused on temperature increase.

To explore and design dynamic qualities of change, cooling has shown to be as important as heating, considering the dynamic qualities when temperature decreases and also its effect during activation cycles.

Besides duration of activation pause, room temperature is also a critical variable in the changes on cooling, particularly in thermal variation of the textile substrate. On heating, room temperature is also more relevant in changes that occur or are influenced by textile thermal expansion, thus affecting with greater impact textile color change considering colorless and color return areas beyond the conductive pattern. Room temperature may or may not be controllable in the context of placement or use of the textile and due to their effect on the textile performances, it is an important consideration in the design process.

The relationship between the textile properties, thermal variation and temporal forms is inherent to all the experiments conducted. The test that has individually approached textile as an independent parameter suggested textile origami morphologies as a design variable to explore dynamic qualities of color change. Furthermore, this study has also highlighted that the first phase of color transition is not merely defined by the integration pattern of conductive materials in the substrates, design possibility also demonstrated in Chap. 4 experiments with the expressive behavior of conductive threads heterogeneous heating.

Dynamic and expressive possibilities analyzed in the experimental studies have built up an understanding on textile behavior, in respect to selected design variables and heat, defining a base framework to further research textiles' temporal color and shape as a physical and immaterial system in interaction with light. Moreover, the studies have also raised new questions for the design research.

Reflection has evolved considering diverse aspects of textile dynamic qualities. Dynamic color and shape changes are not familiar behaviors in textiles and relationships, for example pace of changes being perceived as slow or fast, does not involve previous expectations, such as tree leaves color change during autumn and day and night light cycles. Furthermore, during the experiments analysis it was found crucial to discuss dynamic qualities considering different intensity levels of change, meaning if the textile changes were perceived as obvious or if they were subtle. The understanding of the importance of these expressions of change in how textile behavior is perceived, has raised diverse questions:

What textile qualities may define a subtle or evident change? What pace of change the textiles should perform at to be perceived as subtle? Intensity levels of change are just associated with the speed or duration of change? How is the scale of textile behavior interferes if the change is perceived as evident or almost camouflaged? How can the movement of different textile morphologies address different perception, even performing at the same pace? How can the relationship between textile and dynamic qualities amplify or reduce the intensity levels of change?

With the objective to further research textile behavior and address the research questions, three research prototypes were developed. The objective was to explore and discuss design possibilities of textile color and shape dynamic qualities with different intensity levels of change and to study textile and light interaction. Previous to the prototypes development, an experimental study focused on textiles and light was conducted.

7.2 Textiles in Interaction with Light Transmittance

A series of exercises on textiles and light were developed as a sketching process, an early ideation phase of the research prototypes that explored both fixed and dynamic relationships through physical samples. The main objective was to give insight into *Dynamic Light Filters* qualities.

7.2.1 Textiles and Light Transmittance

Textile morphologies based on origami tessellation techniques involve design decisions that independently of the dynamic behavior, draw different expressive and structural possibilities. Despite the limitless folding techniques and patterns, considerations must be taken into account on how they are materialized. In this research, decisions on textile substrate construction were particularly articulated with the parameters of smart materials integration. However, other folding aspects were explored in respect to the surface stiffness and creasing characteristics.

Textile stiffness can be varied to redefine origami textile expressive and structural qualities. Figure 7.21 shows a selection of samples screen printed with different stiffness agents, folded into a waterbomb pattern and hung in a vertical plane. The first image on the left shows the textile without treatment and the remaining images demonstrate possibilities of how textile stiffness can affect their structural characteristics. In the second sample, there was an evident textile weight increase and low stiffness improvement. The textile held up better the folding structure than the sample without treatment, while presenting a rubber touch. In the third and fourth samples, greater stiffness was conferred to the textile, maintaining the more folded morphology, with a more stable structure in the fourth sample, although the touch was rough.

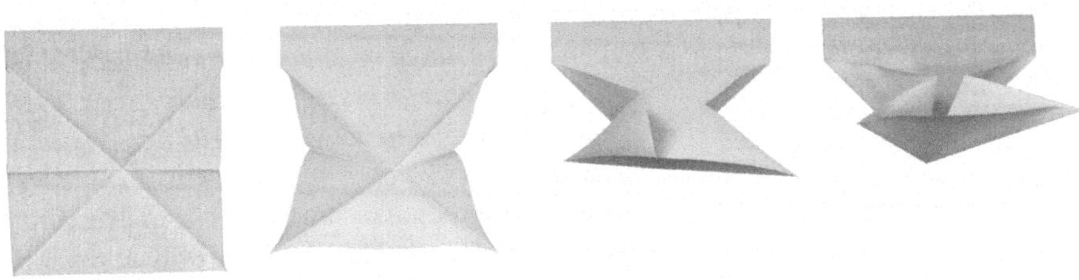

Fig. 7.21 Textile stiffness variation in textile origami

Fig. 7.22 Textile origami with and without creases

Varying the applied agents' concentration, a scale between soft and stiff characteristics can be explored to impart textiles' different qualities. In textile origamis with high dimensions, stiffness is a critical variable as the substrate weight significantly affects the morphologies, thus benefiting of increased structural properties.

Folding processes with textiles can be developed though different techniques such as manual folding with iron creasing; oven pleating with the textile in between two paper sheets previously folded; machine-based pleating processes and sew folding, a technique developed by Rutzky and Palmer (2011) for origami tessellations where similarly to smocking processes, a pattern of dots is sewed and knots are tied to define common points of the textile folded morphology. With these techniques and according to the heat and pressure applied, crease characteristics can vary.

In combination with the textile stiffness properties, decisions concerning the more or less sharp creases influence the textiles and light transmittance expressions. Figure 7.22 presents two squaretwist samples placed above a light box. The textile on the left was ironed, attaining a flat effect and a sharp definition of the geometric light transmittance areas with different layers number. The origami on the right was developed by sew folding without creasing, showing volume in the pile-up layers' areas and rounded edges. In this sample, the layers' distance enables light diffusion in between them, changing how the play of light and shades reveal the textile morphology.

Origami textiles were also explored in respect to changes in light source placement and direction. A squaretwist flat sample was used with the backside turned to the viewer, omitting the central twisted square in the absence of light behind it. Figure 7.23 shows how the textile morphology is revealed without light transmittance and when light behind it is switched on. In the center image, diffused light was used and textile geometric areas with the same layer number attained homogeneous shades. In the right image, the sample was illuminated with side light, placed in the sample top right and pointing to its center. The textile layers present slight thicknesses differences, which were emphasized in side light by the edges brightness. Lighting decisions have composed a pattern of geometric figures and lines, also showing a great variation within darker and brighter nuances. Light placement and direction

Fig. 7.23 Light transmittance patterns

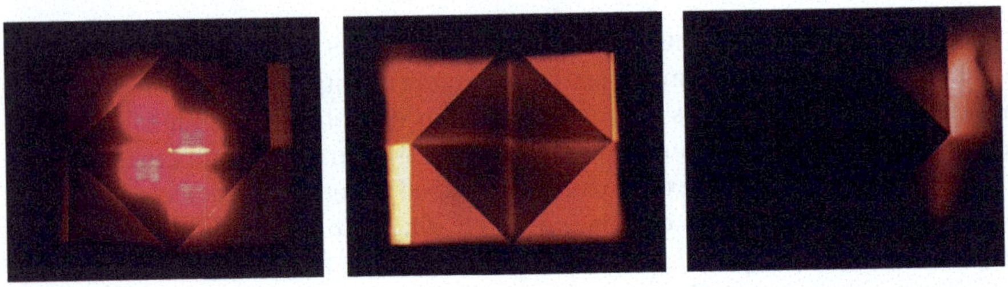

Fig. 7.24 Squaretwist textile above three light structures: LEDs module (left), LEDs stripe (center) and woven POFs (right)

are design variables that can transform the expressions of textile light transmittance patterns.

Artificial light source and placement were defined to be fixed parameters in the research prototypes. A set of samples was developed with the objective to create thin structures that could be placed behind the smart textile and be devised in any shape. Figure 7.24 shows three lighting alternatives produced, recorded with a red squaretwist textile above them. One structure used LED modules set on a rigid base with a diffusion foil in between the LEDs and the textile (left image). Given the textile proximity with the LEDs (2 cm) and their light intensity, a spotty lighting effect was still perceived. LED stripes were set in the laterals of a rigid frame attaining a homogeneous lighting, with a good rendering of the textile morphology (center image). A woven sample integrating side emitting plastic optical fibers (POFs) was also developed and tested (right image), demonstrating very low light intensity, only visible in the textile area of the light-emitting sources. LED stripes option was selected.

7.2.2 *Color, Shape and Light*

Exercises on color, shape and light were developed with the objective to analyze textile behavior in light transmittance and the dynamic effect on light luminosities

Fig. 7.25 Sample Y without and with light transmittance

and tones. In parallel, textile chromic and movement sequences were explored, considering dynamic possibilities with different intensity levels of change, varying the electrical activation parameters.

As an inherent condition of visual perception, light influences how we perceive and experience our surroundings, materials' textures, shapes, colors, etc. Textiles dynamic behavior can transform light that passes through them and light can also transform how textile and textile behavior are perceived. Photos of the textile squaretwist screen printed with TC red (sample Y) illustrate how light reveals the sample colors and morphology, with and without light transmittance (Fig. 7.25).

In the two lighting conditions, the textile coral nuance observed on the left image is transformed in light transmittance, depending on the folded morphology. Textile brightest areas have just one textile layer, increasing color saturation and decreasing color lightness along areas with higher layer number. In result, pile-up textile layers enabled to create strong and dark red expressions, using relatively low pigments' concentration, suggesting the use of light and tessellation origami techniques as additional design tools to explore and create textile colors and patterns.

Furthermore, the way light transmittance reveals textile morphology is not just perceived according to layer variation. The bright horizontal and vertical lines in the central twisted square also reveal how the textile structure overlaps in the samples' backside.

This analysis has pointed out considerations for the prototype development process. When formulating TC pigments pastes for textiles that change color according to predefined color ratios, color comparison and definition should encompass an analysis phase with samples in light transmittance. Moreover, as light transmittance parameters interfere in how colors are perceived, it was important to previously define them. Besides light source and placement, color temperature to apply was defined at this research stage, 2700 K warm white light.

Color change behavior in interaction with light is illustrated in Fig. 7.26. The change from solid color to the color change pattern, gradually allows more light passing through, but the stronger effect is observed with complete areas becoming colorless, creating an enhanced change of light luminosities and tones. The spotty

Fig. 7.26 Color change (1st to 3rd image) and color return (3rd to 5th image) in light transmittance

color change effect occurs in the darkest textile area and the layer shades reduce contrast between the colorless lines and the colorized areas. This dynamic expression is not as evident as when observed without light transmittance, where the colorized dots are more clearly defined by crisp white lines (7.1.1 test E).

Despite the central square being the darkest textile area, it also warms up and becomes colorless faster than the other areas, resulting in slower cool down and thus color return transitions. However, the principle that lighter colors absorb less light cannot be directly applied in this event, since the number of textile layers also affects light absorption, as the 4th image in Fig. 7.26 illustrates. The relationships between color, shape and light are paramount in this research where textile color and shape change are not explored as just two independent behaviors but also how one affects another and enables to extend textile and lighting dynamic qualities and expressive possibilities.

Exercises in color and shape change with samples in light transmittance explored lighting variations and also activation sequences. The experiments conducted took into account the observations and findings attained in Sect. 7.1.1 and projected to gain further understanding of dynamic possibilities through patterns of change.

The concept of pattern in textile design comprises of structural organization of imagery or motifs through repetition, which can be regular, such as straight, half drop and mirror, or irregular (Clarke 2011). Whereas this description is associated to static visual representations, pattern and sense of rhythm evolved through repetitions are also familiar concepts in dynamic events, for example patterns of sounds, movements and cadence of spoken words (Bradley 2012). Exploring patterns of change in textile behavior, the experiments were developed through the repetition of activation and pause cycles with varied durations and electrical current values, to create activation sequences.

In these exercises, the patterns of change defined the patterns of activation considering that activation and pause durations were not previously established, rather they depended on the objectives of each experiment. For example, a color change exercise has studied the possibility to create subtle variations within the dots' diameter. After defining the electrical current value, based on the previous results (Sect. 7.1), power was switched on and off according to the textile chromic behavior. In addition, some exercises involved the variation of the electrical current value, with the objective to study different rhythms of change during the same transition—between one state to another—a possibility previously discussed in test C. With a focus on real-time observation of textile behavior, a video record was kept, but a thermal record was not.

The experimental phase encompassed a comprehensive number of tests, since each exercise included several trial and error attempts. The systematic practice was crucial to develop a better understanding of the possibilities and limitations of thermo-responsive behavior and to explore the relationship between activation parameters through patterns and sequences towards specific dynamic qualities.

The frames presented in Fig. 7.27 illustrate selected activation phases with color change. The first and second frames' rows respect to sections of the exercises on varying the dots' diameter. The discreet changes of color and light required particular attention and tests to select the phase of the dots appearance to switch the power supply off, as in this area the colorless transitions carry on after the activation is stopped. In the first row, 36″ details the moment that power was switched off and until 42″ the colorless area expanded in the central square and color return occurred in the remaining areas, being faster where the textile structure has just one layer. These changes garnered attention for how one activation decision can unfold two opposite color transitions, enabled though the textile morphology.

The second row depicts the changes through repeated activation patterns that are observed particularly when low electrical current values are selected. Differences in textile temperature do not occur merely according to the layer number area, as the center of the twisted square reveals temperature differences, exhibiting small dots that start to blur into colorless areas and the dots near the square perimeter are larger. These two examples demonstrate that exploring subtle textile dynamic expressions can embody intricate performances, which are expressed in a subtle way.

The third row of Fig. 7.27 presents a section of an exercise focused on creating more intense dynamic expression. Sequential changes between colorless and color return around the twisted square attained the effect intended; as more light passes through less textile layers, color change in these areas captured greater light variation. Furthermore, these transitions also displayed a pace pattern through sequential activation, fast on heating and slow on cooling, creating a sense of rhythm in the changes. The paces' differences are illustrated through the textile behavior frames with 10″ interval each, showing greater changes on heating (from the 1st to the 3rd image) and lower changes on cooling (from the 3rd to the 5th image).

Selected sections of shape change sequences are presented in Fig. 7.28. Images in the first row respect to experiments of movement continuity with textile changes displayed after power being switched off. The frames present the textile folding up at 4″, the cooling down with a slow opening movement at 5″, and the activation of the fold down also at 5″. Variations between the activation patterns repeated allowed the changing of paces between fold up and fold down, being: faster/slower; slower/faster; slower/slower; faster/faster. Second row images show a section of the faster/faster experiment.

Faster transitions of the two movements did not encompass activation pauses, either fold up or fold down were in activation stages. High electrical current was selected and through continuous activation patterns of 2″ or 1″, the textile started to display decreasing ability to change shape, given the high temperatures in both Nitinol

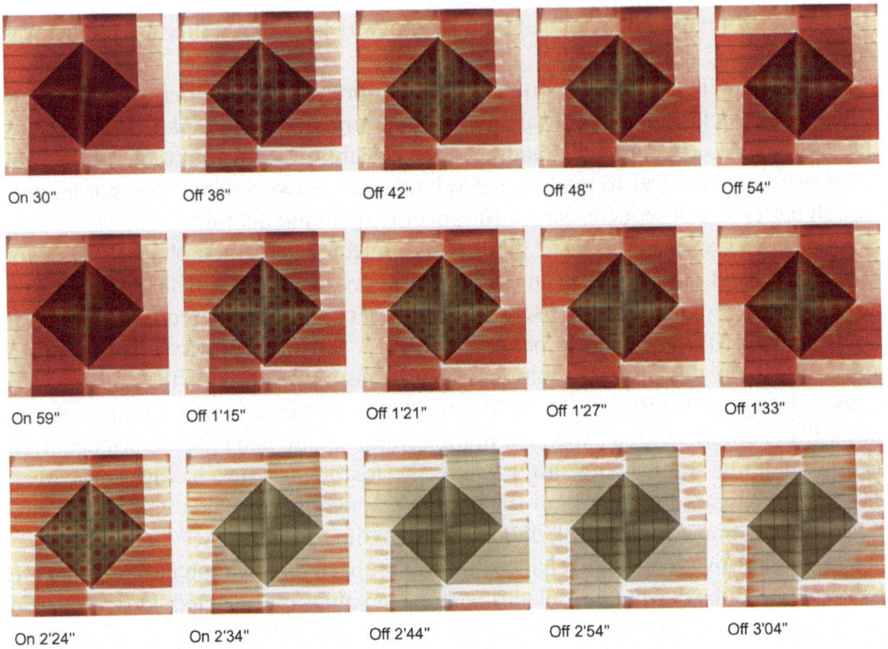

Fig. 7.27 Frames of color change sequences

groups. Whereas this experiment reflects textile behavior limitations, it also adds an expressive possibility to design shape changes that vary from higher amplitudes to lower. To invert these movement transformations, from lower amplitudes to higher, activation pauses were required for the textile cool down, thus increasing the paces of change.

All shape change experiments illustrated in Fig. 7.28 were conducted with the textile sample in the vertical position, but the squaretwist angles are different. The third row experiment was conducted with the textile dynamic tabs diagonal to the horizontal plane. Besides analyzing how this placement parameter interferes in shape change, the possibility to reverse one movement before attaining the final state was explored. Lower row images show the textile folding up (32″ to 35″), starting the fold down movement (35″ to 43″) and reversing again to fold up (43″ to 45″), finally performing the complete fold down (45″ to 58″).

This experiment describes a different method to create a movement transformation similar to the one presented in the second row, but with different paces and greater control to define if the textile performs larger or lower movements, once textile temperatures are lower. Textile morphology displayed a small effect of the squaretwist direction in shape change with the lower textile tab folding up slightly less than the upper tab and reversely for fold down. The influence of the textile direction in this sample was discreet, but with larger and heavier samples, an increased effect is expected.

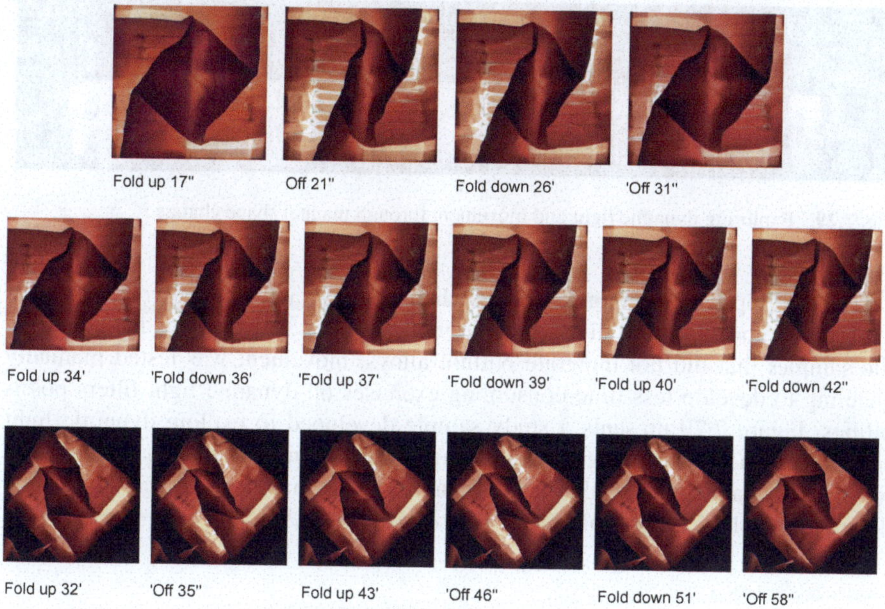

Fig. 7.28 Frames of shape change sequences

Interaction of textile shape change and light is not as distinct as with color change. Subtle lighting transformations are perceived in the sequences illustration mostly by the chromic effect resultant from the Nitinol activation. To research textiles behavior with color and shape change in the same structure, selection of activation temperatures of each material and their thermal variation has to consider not just one behavior, but also the influence that the activation of one behavior can have on the other.

The activation temperatures of the materials selected in this research were 27 °C for TC pigments and 30 °C As—45 °C Af for Nitinol alloys. Shape change cannot be triggered without color change, unless the TC screen printed pattern avoids the Nitinol integration areas, using just conventional pigments. In this case, pattern definition should also consider textile thermal expansion that occurs during shape change activation.

TC trigger temperature is slightly lower than the SMAs selected. Tests conducted with a base sample that integrates Nitinol alloys and conductive threads in the substrate at 5 mm distance each, showed that when power is supplied to the conductive threads, textile temperature increase is able to trigger shape changes. Defining the conductive threads' maximum temperature as 50 °C, through textile thermal expansion the SMAs temperature did not attained the 45 °C required to perform complete shape change and therefore, displayed a small movement and incomplete shape change. The time to heat up the Nitinol alloys depended on the electrical current values and time of activation.

Fig. 7.29 Exploring dynamic light and movement through manual shape change

The development of color and shape change study samples involved a phase of exploring color ratios and diverse textile morphologies. Experiments included textile samples that did not integrate Nitinol alloys, movement was tested manually enabling to develop less time-consuming exercises on dynamic light filters possibilities. Figure 7.29 presents a study sample developed to explore dynamic light expressions and shape change paces through manual transformations. As a method to experiment and generate alternatives, manual shape change was crucial to test and define morphologies and scale of the smart textiles to develop.

7.2.3 Discussion

The exercises displayed and discussed in this section represent a preliminary phase of the prototype development research. Presented in the following sections, the prototypes were not conceived as finished products, they were physical tools to research textile chromic and morphologic behavior as a physical and dynamic system in interaction with light.

The experimental studies developed have presented challenges during photo and video recording. To document dynamic light, camera settings should be constant. In the tests photo recorded, light intensity and tone differences intended to capture have required several attempts to find the parameters that could document the luminous interval. In video, manual settings are not a common parameter in regular cameras, which automatically adjusts the lighting settings. This challenge highlighted the importance of a professional video camera for a more accurate documentation of the research prototypes.

7.3 Case Studies

7.3.1 Research Prototype I: IKEA Museum

The development of research prototype I (Fig. 7.30) aimed to study color and pattern dynamics through a TC electroconductive textile folded with an origami pattern.

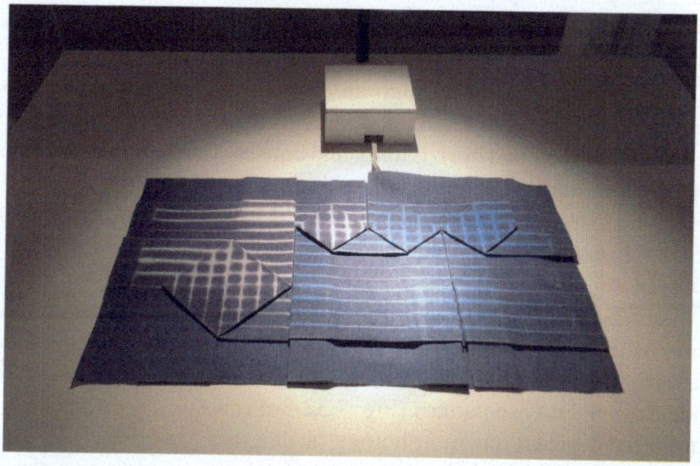

Fig. 7.30 Research prototype I

Fig. 7.31 Color samples with the lower half heated above 27 °C

The definition of the design variables explored took into consideration the experiment on morphology as an independent variable of color change in Test E (Sect. 7.1.1), which analyzed how thermal variation and color change expressions were influenced by the textile structure folded into a squaretwist origami. With this prototype, the focus was on three pattern variables—morphology, heating circuit and color—to explore their relationship and influence on textile dynamic qualities.

The textile color change ratio defined was from similar to different with a temperature increase. A pattern color was set for the TC pigments cold state and three paste recipes were formulated, changing to white, light blue and medium blue upon heating, as depicted in Fig. 7.31 where textile color samples present the upper half at TC pigment cold state and the lower half heated above 27 °C (TC pigments activation temperature). The development of the pastes recipes used to screen print the textile have applied the findings of the instrumental process researched and presented in Sect. 3.4.

Textile morphology was set through an origami crease pattern of squaretwists. The geometric composition defined encompasses three squaretwist base folds with 5 cm square diagonal, aligned horizontally and consecutively, when folded; and 8 cm square diagonal of a squaretwist misaligned in relation to the other squares. The left image on Fig. 7.32 first row shows the origami crease pattern, the textile surface has 53 × 31 cm and image on the right shows the pattern in folded state, totalizing an area of 30 × 18 cm.

After color and crease pattern definition, two parameters were taken into consideration: (a) positioning of the heating lines in relation to the crease pattern—conductive pattern (Fig. 7.32-2nd row); (b) definition of the printed pattern in relation to the conductive and crease pattern (Fig. 7.32-3rd row).

Considering the prototype morphology, a diagram was drawn, which depicts the folded composition of piled textile layers and piled conductive thread layers (Fig. 7.33). The objective of this diagram was to attain a preliminary understanding of the relationship between the patterns previously defined in the textile unfolded state and the geometry that they will assume in the textile folded structure.

The diagram sketching was based on a folded paper model, which allowed to study the spatial composition of the textile and heating lines, previously to the prototype production. By applying a color scheme, the diagram depicts the number of piled conductive threads through lines, in which areas with larger number are expected to attain faster temperature increase; and colored areas that illustrate the textile layer number referring to expected areas with greater textile thermal expansion. Areas and heating lines with lower piled layers are estimated to attain faster color return transitions.

The prototype production is summarized as follows:

(a) Weaving of the textile substrate in a plain weave structure with cotton yarns in the warp and polyester yarns in the weft, where conductive threads were woven at every 1.0 cm.
(b) Formulation of screen printing paste recipes given a pattern color, through a database created with colourimetric properties of the TC and conventional pigments handled; elaboration of the pastes; screen printing and thermo set of the textile.
(c) Connection of the conductive threads in two series circuits totalizing eleven threads each circuit.
(d) Application of an electrical insulating agent in the sample backside and folding the prototype according to the crease pattern previously defined (Fig. 7.34).
(e) Development and testing of the electrical circuit (Fig. 7.34).

Initial analysis of prototype behavior was conducted with a variable power supply, at a room temperature of approximately 20 °C. Images displayed on Fig. 7.35 show one cycle of textile color change activation from colorized to decolorized state. Previous to activation, the prototype is monochromatic and its pattern is defined through geometric morphology. At the first stage of activation, the textile changes

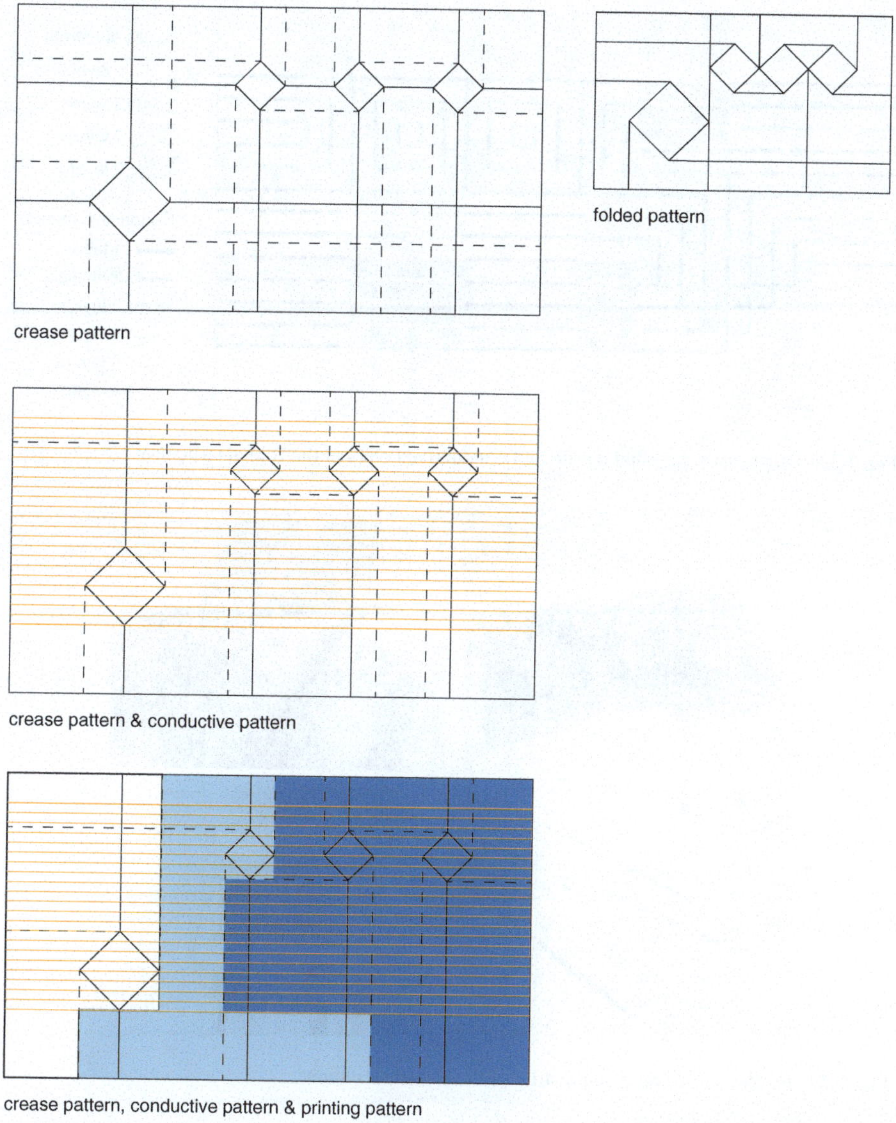

Fig. 7.32 Prototype I crease pattern, folded pattern, conductive pattern and printing pattern

from solid color to a striped color change expression, displaying the pattern of the conductive threads within the folded structure. Through textile thermal expansion the linear stripes slowly give rise to complete areas of color change. During this transition, awareness of the color change ratio increases, becoming more evident the variation from one initial color to four, namely three colors at the TC heated state and the cold state color. Along this stage, textile expression builds up on morphology, conductive

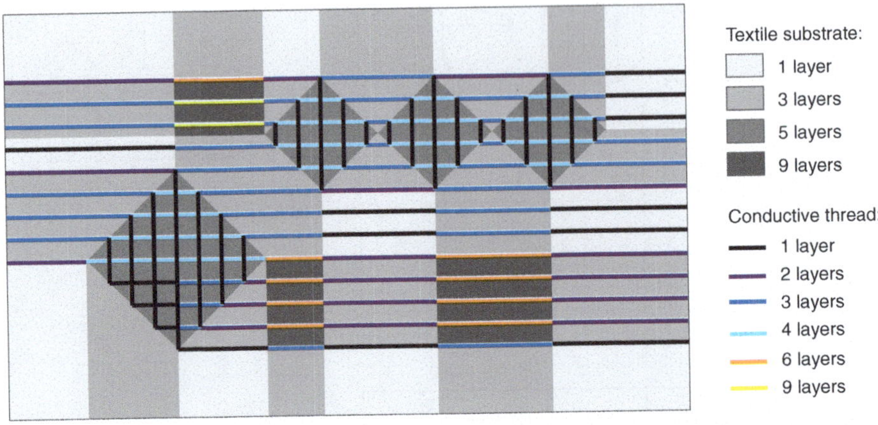

Fig. 7.33 Diagram of the piled textile and conductive layers in the folded structure

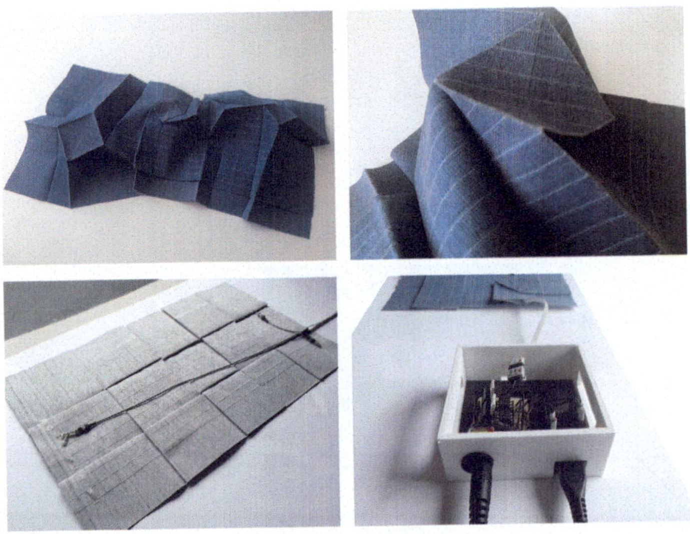

Fig. 7.34 Prototype I folding and testing of the electrical circuit

and printed pattern, as the linear form of the heating elements are still detected at the same time that areas of color differences start to reveal. When the dynamic behavior attains complete areas of color change, the conductive lines becomes blurred and the textile pattern solely encompasses morphology and printing variables.

Colorization pace depend both on textile and conductive layers, whereas the number of piled heating lines has higher impact. The squaretwist that changes to white has equal number of textile layers but different conductive layers. For example, the lower left section piles up two conductive layers above three non-conductive textile layers (one with horizontal heating lines and one vertical lines), while the upper half

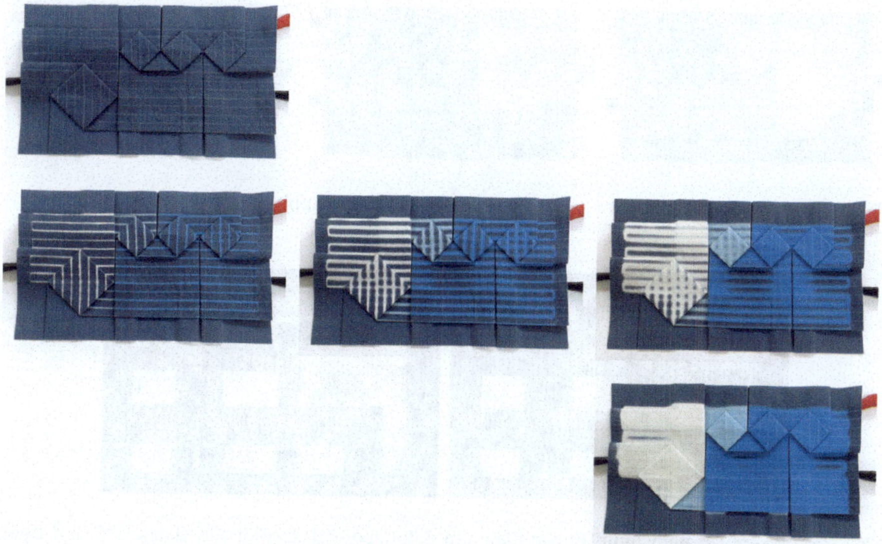

Fig. 7.35 Color change transition from colorized to colorized state, through electrical activation

Fig. 7.36 Prototype section in color change

totalizes five conductive layers (four with horizontal lines and one vertical lines). As a consequence, pace of color dynamics varies, enabling to design heterogeneous expressions of the spotty color change effect (Fig. 7.36).

The influence of layer number also applies for the transition from decolorized to colorized state. Figure 7.37 shows images recorded during color return, after the prototype have attained complete areas of color change. Slower paces occurred in the sections with higher piled layers, both textile and conductive, but is was also observed their effect in nearby areas, as perceived in the upper right square that changes to white. Textile color return depends on temperatures previously attained

Fig. 7.37 Color return transition, after interruption of power supply

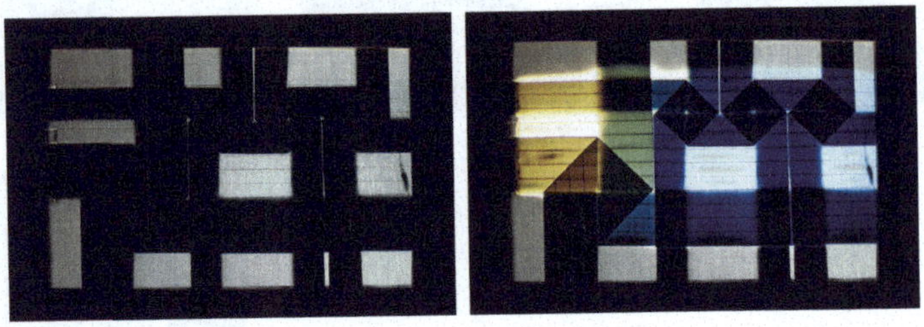

Fig. 7.38 Prototype with light transmittance in colorized and decolorized states (left and right, respectively)

considering areas with different textile and conductive layers, as well as temperature of adjacent areas.

The prototype behavior was also analyzed in interaction with light. Figure 7.38 shows the textile at colorized and decolorized state (left and right images, respectively). At cold state, the light transmittance pattern emphasis the textile folded morphology with distinct luminosity areas. When heated, a colored pattern emerges and relationship between pattern variables is more evident.

The squaretwist that in the square surface changes to white is heterogeneous in the number of piled conductive layers as well as the piled layer colors. When the prototype activation attains complete areas of color change, four chromatics can be distinguished in this square area, namely white, aqua, blue and black. In this sense, the dynamic light filter effect reveals in a more intricate fashion the relationship between patterns—morphology, conductive and printing—than the textile color change without light transmittance.

Besides textile expressions being dependent on textile layer number and piled conductive layers, they are also influenced according to color and particularly lightness dimension. The three squaretwists aligned change color to medium blue and their square area presents five piled conductive layers. Comparing with the upper half of the larger squaretwist that has the same conductive layer number and textile layers in white and light blue colors, their luminosity is distinct, with the medium blue squares being perceived as black. In addition, the linear pattern of the conductive circuit is also more visible in areas with lighter colors.

The prototype demonstrates that the light transmittance pattern expresses the relationship between morphology and colors built upon layers, particularly when the textile is in a heated state.

The prototype was presented at a temporary exhibition of Ikea Museum in Älmhult, Sweden, from 3 June 2017 to 2 April 2018. The exhibition entitled 'Textile

Fig. 7.39 Prototype presentation

Playground' and curated by Raphael Bartke, invited visitors to explore Ikea textile history and to play with patterns, also giving insights into emergent research and developments in the Textile Future section, where the research prototype I was presented (Fig. 7.39).

An electronic circuit was developed to program the textile color change activation. Visitors could initiate the activation sequence by pushing an electronic switch. The sequence comprehended a phase with sequential and simultaneous activation of the two series circuits as well as a period where power is off, so the textile temperature could decrease, even if consecutive activation cycles were triggered.

The time off programmed did not intended to necessarily cool down the textile until room temperature (approximately 20 °C). In this sense, visitors influenced the color change expressions in respect to the textile activation history, before initiating a new activation cycle.

The prototype behavior displayed visual performances of textile color and pattern as a result on materials and design research towards exploration of textile new perspectives and application.

Development and presentation of the research prototype allowed to study color change behavior with focus on textile pattern within three variables: crease pattern, conductive pattern and printing pattern. Whereas they can be considered as individual parameters, their definition builds up upon each other creating a framework for the textile performance.

Differences on the textile and conductive layer number present a visually perceptible effect on the color change paces and expressions, either on heating and on cooling. With the color change ratio applied—from similar to different—printed pattern is perceived on heating, according to textile morphology and lighting conditions. Without light transmittance, color change is observed on the surface of the folded textile, while in light transmittance the chromatic compositions depend on the number and colors of the textile layers, adding complexity to colored pattern.

In addition, color change behavior is also dependent on materials properties, integration processes applied and influenced by design decision on color, color change ratio and activation sequence. In this prototype presentation, visitors also played an active role on textile expressions over time.

The schematic created to analyze the described parameters in the folded structure (Fig. 7.33), before the prototype materialization, provided an understanding of the spatial relationship between textile, conductive and pattern layers and allowed to pre-establish a link with the potential dynamic qualities, in respect to temperature variation and textile thermal expansion in different areas of the prototype. This method can be applied during the design process of thermo-responsive textiles, assisting decisions for the textile physical definition and creation of temporal expressions.

7.3.2 *Research Prototype II: Smart Textiles Salon*

Narratives of Winter Daylight is a research prototype inspired by the dynamic qualities of natural light. Day and night commonly present different rates and changing patterns of light intensities, colors tones and direction, depending on the geographic location, season and weather conditions. For this research, one winter day was observed and photos were recorded each hour from 6 a.m. to 6 p.m. Figure 7.40 presents the photo record conducted on 31.12.2015 at latitude: 41.178084, longitude: −8.66061, with equal camera settings for a comparison between luminosities (Cabral et al. 2017).

Perception of natural light variations involves a relation between the luminosities observed with established expectations and rates of change (Cuttle 2015). Commonly, subtle light changes are expected during the day, while more obvious changes are prevalent in early morning and late afternoon. Rhythms of light variation can allow an experience of the passing of time, perceived not only by the changes of the light parameters but also in respect to the duration of the events in time (Le Poidevin 2015).

Fig. 7.40 Photos record from 6 a.m. to 6 p.m. (Cabral et al. 2017)

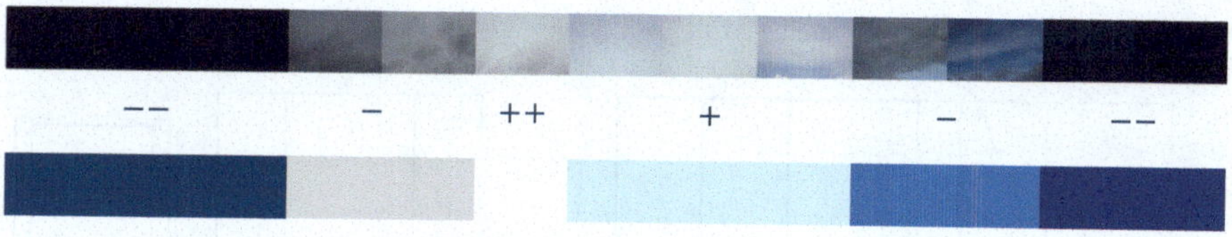

Fig. 7.41 Conceptual bar of luminosities and colors

During the day observed, higher luminosity levels were perceived from 11 a.m. to 2 p.m. with predominant light greys and light blues. From 9 to 10 a.m. the sky tones varied between cold greys and from 3 to 4 p.m. it changed between blues. Before and after the time periods described, light intensities were reduced regarding the camera fixed parameters, resulting in dark images. Camera settings were tested in the previous days to the record presented, but a higher range of light intensities was not attained without blurring the photos in the brighter day periods. By direct observation, a cold greyish dark blue was observed in early morning and in late afternoon a subtle violet nuance in a dark blue.

Along the paces of light variations illustrated through the photos, faster from 8 to 9 a.m. and from 4 to 5 p.m., clouds also created changes during the brighter periods. Although the luminosity changes were low, compared to the overall day and night cycle, obvious changes were created, occurring in shorter time periods. The relationship between patterns and expressions of light change with perceived luminosities and tones, set up a basis to define the prototype color change palette and shape changing morphologies, having also delineated dynamic qualities to explore in the prototype behavior. Initially, the photos were set in one row and related luminosities and colors were interpreted and grouped according to similarities between time intervals (Fig. 7.41). Symbols '–' and '+' were used to establish a hierarchy of luminosities.

With the objective to explore intensity levels of change according to color differences, textile color change defined comprises of variations between similar colors, different blue tones and different colors. The color change ratio defined for the prototype was from similar to different upon heating, with a dark blue color in the cold state changing to the range of colors presented in the bar, when above the activation temperature of the TC pigments handled. Furthermore, it was intended that shape change activation would not change the textile color and a screen printing paste that combined just conventional pigments was defined for the Nitinol integration areas.

To study how different scales of shape change interfere in the dynamic expressions, textile transformation aimed to create volume variations in individual textile areas with different dimensions. Dynamic morphologies were explored through folding techniques in particular through origami pleats, consisting of parallel fold lines, alternating valley and mountain creases, or the inverse (Jackson 2011).

The prototype morphology presents two piled up textile layers. The front layer combines six pleat units aligned horizontally in the textile substrate and the geometry defined for each unit consists of a mirrored sequence of two vertical pleat folds (Fig. 7.42 left). Attached to a plain textile layer (back layer), the folded geometry

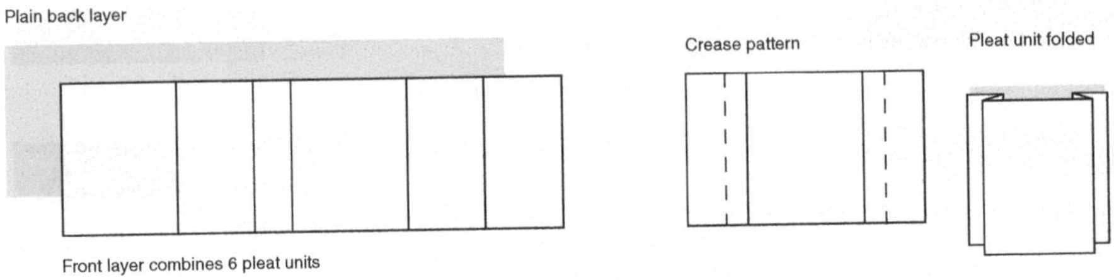

Fig. 7.42 Prototype layout folded (left), pleat fold crease pattern and pleat unit folded (right)

creates areas with a variation of textile layer number and distance, through the fold up and fold down movements of the pleats (Fig. 7.42 right). Width proportion of each pleat was defined according to the luminosity and color similarities set framework.

Textile morphology and colors were tested in light transmittance with study samples. Shape change with textile pleat units in light transmittance included analysis through manual transformations, as depicted in Fig. 7.29 of the previous section. The prototype intended to explore calm and smooth transitions of textile color and shape states, to transform light transmittance through vary dynamic expressions. It was defined that each pleat unit could be individually activated for fold up and fold down and display color change in the front and back layers. Thus, the six pleats' prototype totalizes 24 activation channels.

In addition, the dynamic light filter behavior focused on changes in local lighting, proposing a closer spatial relationship with the textile. The prototype scale was defined according to the possibility to observe the changes in proximity and dimensions when folded were approximately 70 × 24 cm.

The integration processes of smart and conductive materials in textile substrates previously researched and discussed in Chaps. 3, 4 and 6 were implemented in the prototype development. A description of the processes conducted is summarized as follows:

(a) Definition of textile shapes and respective weaving plan for Nitinol wires integration; definition of Nitinol geometries encompassing two sets of Nitinol wires (one group to perform fold up and another for fold down); drawing and production of dies; optimization of the annealing parameters followed by Nitinol heat treatment; and sandblasting of Nitinol wires' ends.

(b) Weaving of the textile substrate in a plain weave structure with cotton yarns in the warp and polyester yarns in the weft, where conductive threads were woven at every 1.0 cm and the Nitinol alloys were integrated at every 2.0 cm, with 0.5 cm distance of the conductive threads.

(c) Formulation of screen printing paste recipes given a pattern color, through a database created with colourimetric properties of the TC and conventional pigments handled; elaboration of the pastes; screen printing of the textile substrate with TC pastes and after with finishing agents to increase stiffness (according to methods described in Fig. 4.1) in front areas of each pleat; and thermo setting (Fig. 7.43).

Fig. 7.43 Prototype fabrication and testing of the electrical circuit

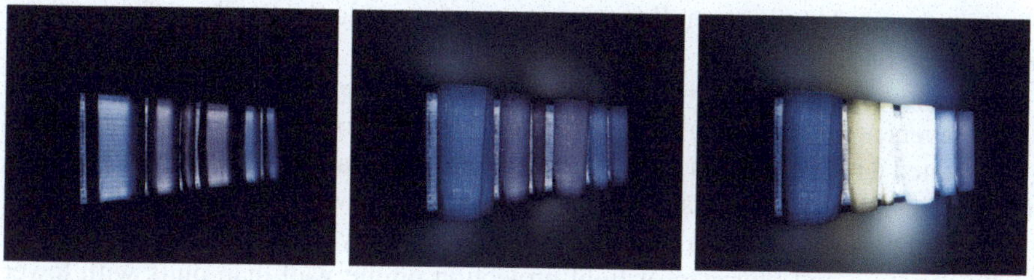

Fig. 7.44 Fold down to fold up in colorized state (1st to 2nd image) and color change in fold up state (2nd to 3rd image)

(d) Connection of the conductive materials with four circuits for each pleat unit—two series circuits of connected conductive threads, one for color change in the front layer and another for the back; and two series circuits of Nitinol segments linked according to fold up and fold down groups.
(e) Application of an electrical insulating agent; folding of the prototype and setting up in a light box, built with two strips of warm white LEDS (2700 K) in the box inner top and bottom.
(f) Definition and manufacturing of the electrical circuit and programming the textile activation (Fig. 7.43).

The prototype dynamic behavior was analyzed with the objective of studying relationships between temporal forms—color, shape and light—and the expressions of change. A first phase of activation tests has examined the resistive heating properties of the conductive channels one at a time, using a variable power supply, a multimeter and a thermal camera. After, a temporary electrical circuit was developed, linking the 24 conductive channels to an Arduino microcontroller, so the activation could be performed and programmed with more than one heating element at a time. Finally, the prototype was placed on the wall and the analysis included direct observations, photo and video records.

Figure 7.44 presents the prototype in three different states. From left to right images, the textile pleats change from fold down to fold up in colorized state and then decolorized in the fold up morphology. Through the simultaneously activation

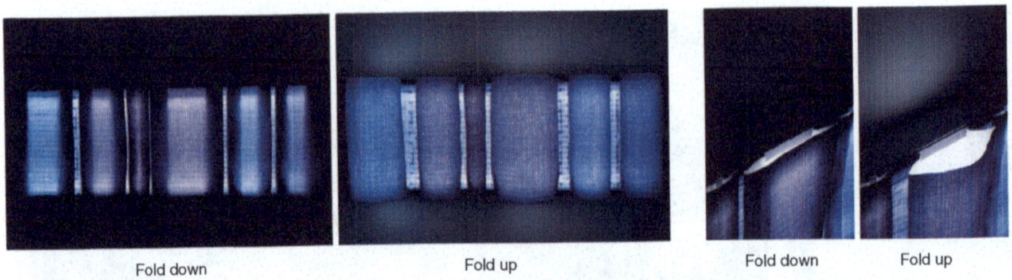

Fig. 7.45 Prototype fold down and fold up activation (Cabral et al. 2017)

of all pleats, color change created higher light transmittance variation than shape change, observed by the overall luminosities and tones in the textile surface and the light falling on the wall.

With the fold down activation, light reflection on the wall is imperceptible. The textile layers are near and light transmittance is low in comparison to the morphologies attained with fold up, where the pleats unfold and create 'open pockets' (Fig. 7.45 right images). During this change, pleat units with larger width (Fig. 7.45 left images, 1st and 4th pleats) attained a larger distance between layers, achieving higher luminosity variation and blur size of light projection on the wall than units with lower widths.

When the prototype was observed at different viewing angles, shape change expressions also differed. This relationship was analyzed through Kleins' shape transformation types (Nefs 2008). When observed at a frontal view, change of shape produced through the pleats fold down and fold up can be defined as dilation, a similarity change where shape angles and distance ratios are maintained. The textile movement was perceived as very subtle, even at the faster pace with the larger width pleats. It was light transmittance variation that appeared to create a noticeable dynamic expression.

At a side or perspective view, the movement was clearer. The shape transformation was observed with different distance ratios—strain transformation—enabling a greater awareness of the shape change as well as the relation between the textile and light variation.

The importance of textile movement in the intensity levels of change is demonstrated through different viewing angles. This relation identifies morphology, transformation type and positioning as design variables in the dynamic qualities and expressions of textile behavior.

Through the shape change analysis, it was observed that folded states attained lower angles than the memorized geometries of the Nitinol alloys. Textile substrate and morphology have constrained the Shape Memory Effect (SME), in particular with the lower width pleat, where movements were barely perceived although the Nitinol wires were above their activation temperature, as confirmed with a thermal camera.

Fold up and fold down paces were analyzed through electrical current variation. At a room temperature of approximately 20 °C, mean values attained for each circuit were 14 s with 0.6 A, 10 s with 0.7 A, 7 s with 0.8 A and 5 s with 0.9 A. Movements performed appeared organic along the different paces. Although, as demonstrated in test F of Sect. 7.2, the final activation phase with the lower electrical current values is barely perceived by direct observation.

For a comparison between the scale of change and different paces, a larger pleat unit was programmed to fold up more slowly than a medium pleat unit. The changes with the larger pleat were perceived to last slightly long, but the movements were calm in both pleats. The scale of change from one state to another appeared to be small for a comparison between paces of the textile morphology. Instead, consecutive cycles of fold down and fold up movements appeared to enhance the differences. Movements in the medium pleat were perceived as faster and the larger pleat accomplished slower changes but with higher movement amplitudes. The change of the movement direction added a new variable in the analysis, emphasizing the differences of the movements' qualities.

Through the sequential shape change activation, the actuators temperature increased faster in particular in the pleat activated with high current values for faster movements and few cycles were performed until an activation pause was required for cool down. Time for the Nitinol to cool down was tested. As previously discussed, the alloys temperature is not homogeneous and their cool down depends on several variables such as previous activation time, temperatures attained with the electrical current supplied and textile substrate characteristics. With the prototype activated at a room temperature of 20 °C, values varied between 5 and 20 s.

The relationship of shape change and light also depended on the textile color state and dynamic qualities of chromic behavior. At the fold up morphology, textile color change in all pleats is depicted in Fig. 7.46. Light transmittance variation with the color change effect, created a hierarchy of luminosity differences as targeted: an increase from the first to the third pleat (6–12 a.m.); similarities in third and fourth pleat (12–2 p.m.); followed a decrease until the sixth pleat (3–6 p.m.). Light measurements with one layer of each color sample confirmed the observations, percentual luminosity differences varied from 13% to 76, 93, 88, 58 and 30%, in respect to the above-mentioned sequence.

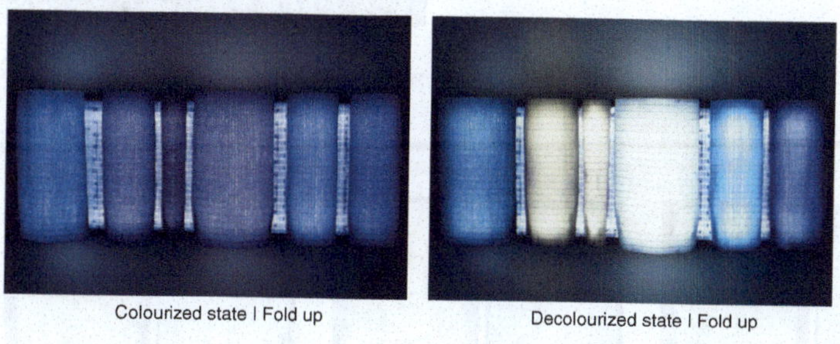

Fig. 7.46 Prototype without and with color change activation in fold up state (left and right, respectively) (Cabral et al. 2017)

Color tones perceived by direct observation of textile samples without light transmittance and with the textile placed in a light box have differed (Fig. 7.47). Color measurements conducted below 27 °C in a reflectance spectrophotometer have confirmed color similarities. CIELAB differences attained values between 0.8 and 1.3 dE*, which can be considered reasonable for color matching. In the light box, samples screen printed with pastes that combined TC pigment blue (2nd, 3rd and 4th pleat units) instead of conventional pigment blue, presented a violet nuance. Paste recipe adjustments were conducted, although a subtle violet tone remained (Cabral et al. 2017).

From similar to different, color changes varied light transmittance heterogeneously and affected the luminous relationships attained with shape change. For example, shape and light changes were more evident with pleats of larger width (1st and 4th) in colorized state (Fig. 7.47 left image). In colorless state it was observed that luminous changes with lighter colors pleats were more obvious, even with smaller pleats (Fig. 7.47 right image). Reversibly, textile morphology also affected how colors were perceived. The smaller pleat (3rd) changed to white, but due to the lower textile layer distance, light shadows made it appear darker than the light blue pleat (4rd), thus establishing a relation between the scale of color change, morphology and light transmittance variations.

Paces of color change were analyzed in all pleats. At a room temperature of 20 °C and with 1.2 A, mean values to attain full color change in the front textile layer activation was approximately 2′30″ and 1′ for color return. Through simultaneous activation of the front and back layers the change to colorless was faster (2′) and color return slower (1′20″). The temperature inside the folded layers increased, enhancing the textile thermal expansion and, due to the vertical positioning of the prototype, the upper area of the pleat units also heated up faster than in the bottom (Fig. 7.48)

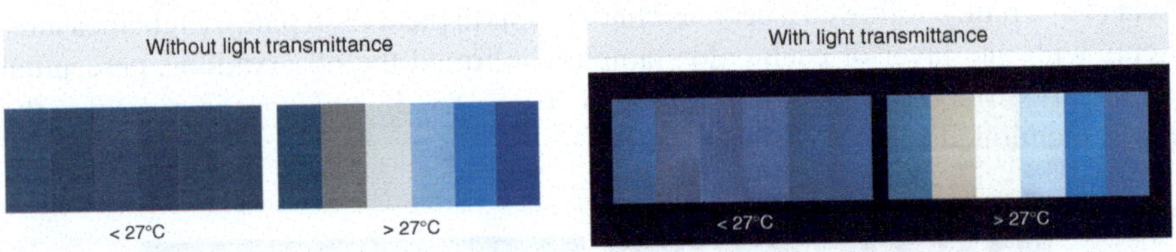

Fig. 7.47 Color samples below and above 27 °C, without and with light transmittance (Cabral et al. 2017)

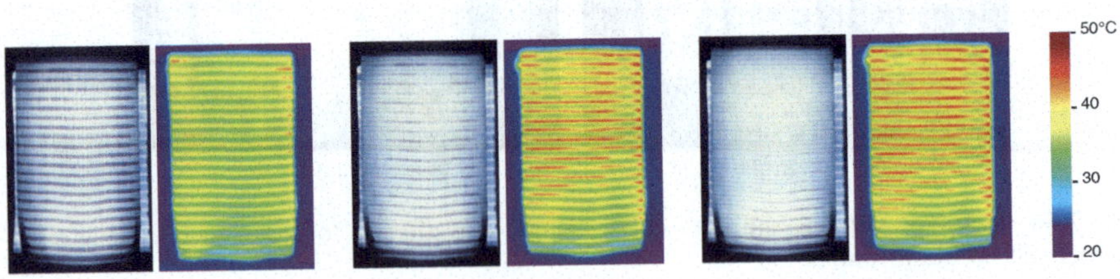

Fig. 7.48 Photos and IR images during color change

(Cabral et al. 2017). With both layers activation, time variation in heating and on cooling is emphasized when consecutive cycles are run, also changing the textile expressions in particular with less sharp definition of the color change stripes.

The rise of temperature in between the textile layers can also interfere with shape change activation and even induce changes in the fold up and fold down Nitinol groups without power supply, assuming a morphology between states (Cabral et al. 2017). It was observed that these changes were more evident in the upper area of the pleats, also resultant from the vertical position. Color and shape analyses highlighted the importance of the relation between textile morphology and positioning in the dynamic qualities of thermo-responsive textiles.

Color change in light transmittance was also investigated in respect to color differences in the colorless state. From colorized to the striped pattern, changes between hues are not necessarily more obvious than changes between color tones. Color lightness appeared to have a more relevant impact in the intensity levels of how the changes were perceived. From the linear pattern to colorless areas, textile colors are more clearly distinguished and also reflected in the tone of light on the wall, also according to textile morphology.

The prototype was presented and analyzed in a spatial context, exploring color, shape and light as dynamic variables of responsive environments. The objective was to display expressive possibilities of change through the textile behavior and dynamic local lighting and create visual Narratives of Winter Daylight. Considering the day observed, the rhythms studied and programmed focused on predominant calm expressions and punctual more intense changes.

An interaction diagram was defined, establishing a relation between the spectator(s) and the dynamic behavior of the prototype. As illustrated in Fig. 7.49, three ultrasonic sensors were used to detect the presence of the spectators near the prototype. The sensors were connected to a microcontroller and transmitter to communicate via wireless with the receiver and microcontroller set, connected to the prototype. According to the visitor's positioning and proximity to the textile, sensors' combination input defined a programmed sequence that activated the textile behavior.

The definition of activation sequences built in the previous studies conducted with the prototype and focused on textile and light dynamic qualities expressed in subtle ways. To depict the textile activation within the 24 conductive channels and guide the respective programming, a graphical setup was developed. Figure 7.50 presents an activation sequence setup that displays a repeated pattern of change from left to right pleats with a delay of 40″ between them. Color change in the front layer starts in fold down state and in the last phase of the textile thermal expansion, back layer is also activated and until full colorless, the textile also folds up. On cooling, gradual color return occurs approximately during 1′ in each pleat. The defined sequence explored smooth transitions between lower and higher luminosities in each pleat and lower to higher and lower again in the pleats sequence.

The prototype presentation took place at Smart Textiles Salon 2016 in Ronse, Belgium (Fig. 7.51). The interactive and dynamic scenario built up through the prototype performance was perceived as slow and calm. Within mostly discreet changes, color

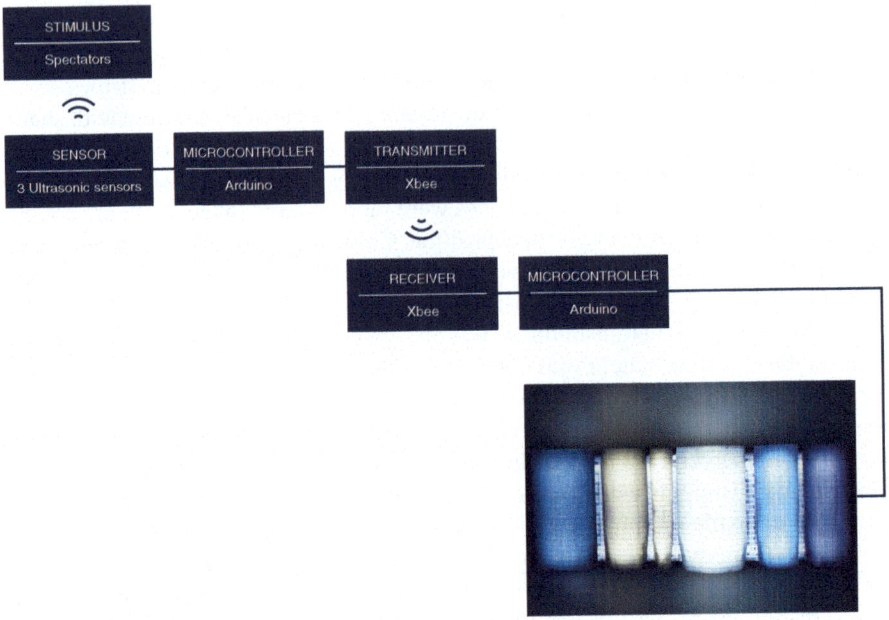

Fig. 7.49 Interaction diagram

change in pleats with lighter colors in colorless areas enabled to vary the intensity levels of the changes perception, in particular through the linear pattern of color change, as well as light transmittance variation with larger width pleats when completing the chromic behavior. Nevertheless, the slow paces defined have still conferred a calm expression to the more evident perceived changes.

Moreover, the active role of the visitors on the textile activation was also discreet. Depending on the proximity with the prototype and positioning, an activation sequence was triggered. Although, the subtle changes did not become obvious at the beginning or the end of the interaction, they appeared to gradually blur within the still phases. The organic movements and subtle rhythms defined a statement of the textile behavior and light expressions.

Designing smart textiles in interaction with light that express subtle changes, reveal possibilities to seamlessly integrate the tangible and immaterial aspects of the smart textiles in the environment, their rhythms and contexts of use. This relation draws on the line of calm technology, a concept introduced by Mark Weiser and John Seeley Brown in 1995, which describes possibilities of information technology moving between the periphery and the center of our attention (Weiser and Brown 1995; Benitez 2016). When designing textile behavior and light, dynamic qualities of smart textiles can be explored as an integral and unobtrusive element of the environment, in which functional and expressive behavior evolves without necessarily engaging or requiring attention.

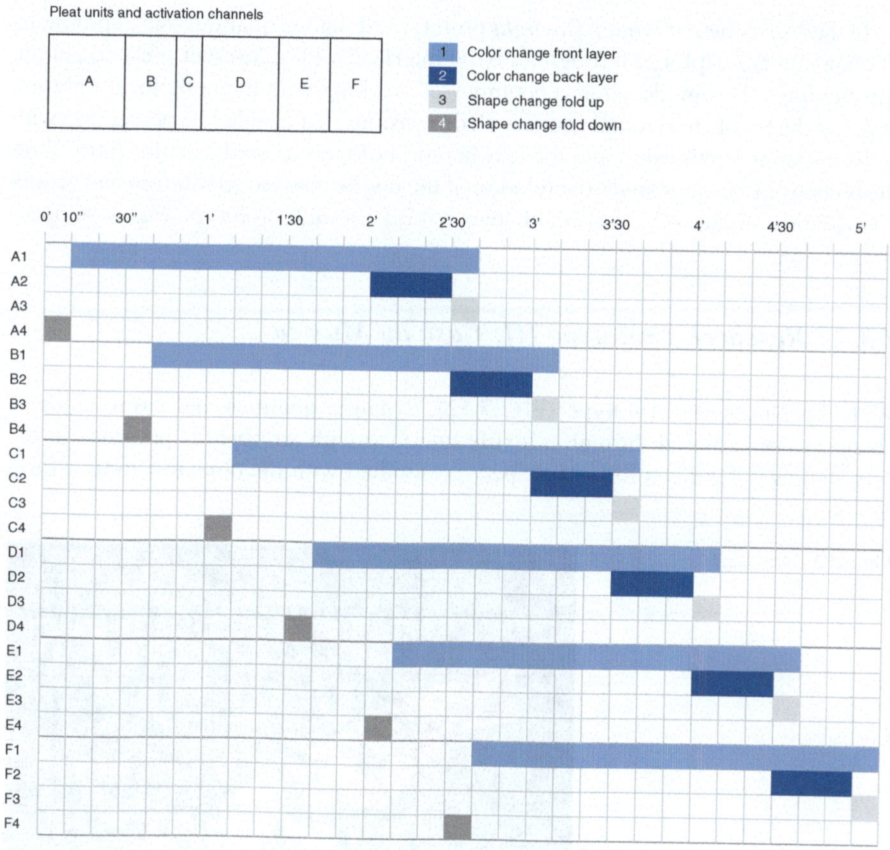

Fig. 7.50 Activation sequence setup

Fig. 7.51 Prototype II presentation

In the *Narratives of Winter Daylight* prototype, dynamic qualities and expressions of change were explored through a textile interface with a discreet presence in the surroundings. Design decisions encompassed working with dynamic local lighting, defining the textile horizontal frame with dimensions that enabled its perception with a close spatial relationship and the textile morphologies as well as color ratio. With the objective to further study the relation of the textile interface definition and design possibilities of intensity levels of change, a third research prototype was developed.

7.3.3 Research Prototype III: Casa da Música

With the *Dialogues* prototype (Fig. 7.52), dynamic qualities and expressions of change were explored through a textile interface with an obvious presence in the surroundings. Textile color and shape were studied with the objective to transform

Fig. 7.52 Research prototype III

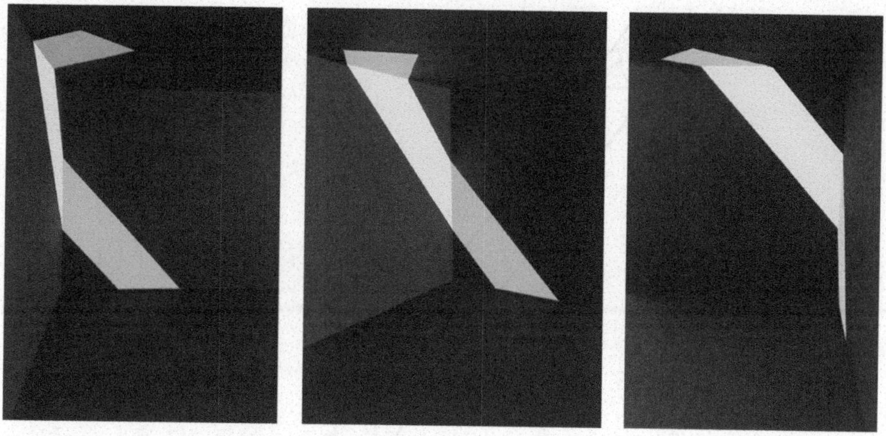

Fig. 7.53 Prototype geometric layout

ambient light. A natural light and color phenomena commonly observed in the northern and southern hemispheres have inspired the prototype development: the polar lights, also named auroras.

Aurora borealis (north) or aurora Australis (south) are night skylights that display color and pattern changes, as a result of chemical and physical interactions in the atmosphere. Visual manifestations of the auroras involve possibilities to vary between forms, such as 'curtains' or 'arcs', display intense or dim luminosities, perform rapid modification or steady glows and change between different colors, according to the altitude that the interactions occur. From lower to higher altitudes, colors of light are blue and violets; greens; and ruby reds (Yahnin et al. 1997; Hall et al. 2002).

The parameters identified in the colorful skylight phenomena established a framework to explore textile color and shape behavior in interaction with light. Considerations included the relation of colors with altitude and changes of form within the color pattern; scale, direction and rhythms of change within varying light intensities.

Focusing on dynamic ambient light in interior spaces, textile morphology studies included folded surface pattern and the layout of the textile structure. The textile layout explored connections between scale, positioning and viewpoint possibilities. The prototype was defined with three textile modules that compose a geometric configuration in two adjacent vertical planes of an interior space and one horizontal plane (Fig. 7.53).

Each textile module was defined with an intricate pattern of squaretwists. The selection of this origami tessellation base-fold took into consideration the morphology transformation type through the direction shift of the square tabs folding up and down; the color effects created through the textile layered morphology in light transmittance. Figure 7.54 depicts the origami morphology in folded state with the three modules aligned, distinguished through the thicker lines. From the prototype bottom to top, squaretwist dimensions decrease and the direction of the folded pattern varies among modules 1 and 2 (M1 and M2), as depicted through the grey shades. The

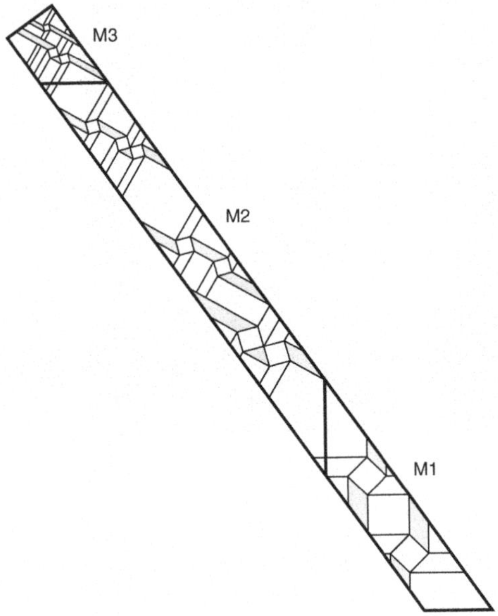

Fig. 7.54 Prototype folded pattern

shades also define the squaretwist tabs that perform shape change. M1 and M2 will be placed in vertical positioning and module 3 (M3) in horizontal.

Nitinol alloys integration in the woven structure and respective geometries were defined through the origami crease pattern of each module and respective shape change set. Figure 7.55 presents M3 crease pattern with Nitinol fold up group depicted with light green lines and fold down group with dark green. To apply the origami crease pattern in the weaving plan, particular attention was required in respect to the textile shape change direction. Considering the alloys integration in the weft, the axis of the moving tabs required to be in vertical position in the weaving plan and given the loom 150 cm width, the rotated crease pattern had to fit within this dimension.

With the objective to explore dynamic colored light and changes through luminous nuances, textile color ratio defined was from different to similar with temperature increase. The relation of auroras' colors according to altitude inspired the studies of textile color and screen printed pattern. Color experiments included analysis of folded samples in light transmittance. Figure 7.56 displays a set of study samples developed comprising violet, blue, green and red colors below the TC pigment activation temperature (left image) and yellow after heating (right image).

The screen printing pattern was defined with the surface in unfolded state, where geometric areas of solid color were delineated along the flat structure. When the surface was folded, the origami morphology created a new geometric composition that presented areas of piled up layers with more than one color, thus slightly changing tones when revealed in light transmittance.

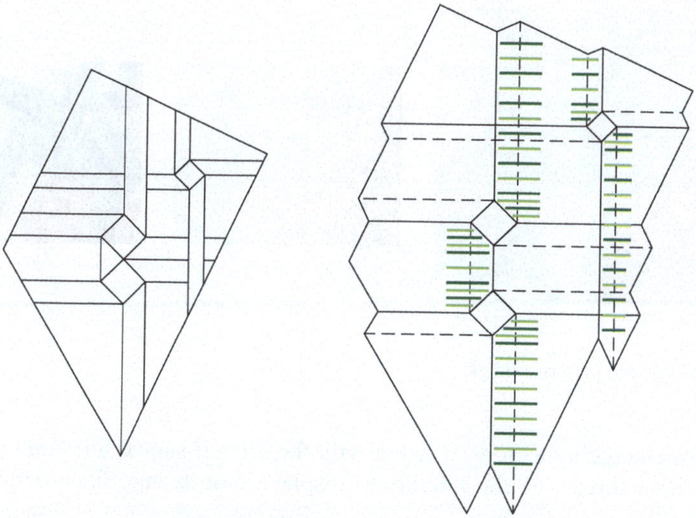

Fig. 7.55 M3 folded pattern (left) and crease pattern (right)

Fig. 7.56 Color studies with folded samples in light transmittance, below and above 27 °C (left and right, respectively)

Figure 7.57 presents a section of the textile pattern unfolded (left), folded (center) and as revealed in light transmittance (right).

The prototype activation was defined with 45 channels. For shape change, 2 squaretwist pairs are activated simultaneously and the remaining 8 squaretwist individually, totalizing 20 channels (10 fold up and 10 fold down). For color change, the conductive threads were connected in series into 25 circuits. In some areas, the connections were carried out in consecutive threads and other areas with intercalated threads. This decision took in consideration the possibility to create gradual color

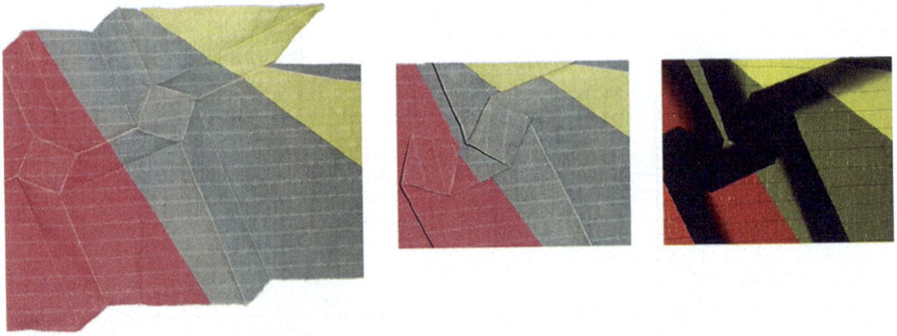

Fig. 7.57 Color and pattern studies

and pattern change transitions, starting with the 2 cm distance lines and activating the conductive threads in the middle, to display color change lines with different thicknesses.

The prototype production is summarized as follows:

(a) Definition of textile shapes and respective weaving plan for Nitinol wires integration; definition of Nitinol geometries; drawing and production of dies; optimization of the annealing parameters followed by Nitinol heat treatment; and sandblasting of Nitinol wires' ends.
(b) Weaving of the textile substrate in a plain weave structure with cotton yarns in the warp and polyester yarns in the weft, with Nitinol integration according to the crease pattern defined for each module.
(c) Elaboration of the pastes, according to the color samples selected with constant concentration of the conventional pigment yellow (1.5%); screen printing of the textile substrate and thermo setting (Fig. 7.58).
(d) Connection of the conductive materials according to the defined layout for color and shape change (Fig. 7.58).
(e) Application of an electrical insulating agent; folding of the prototype and setting up in a light box, built with two strips of warm white LEDS (2700 K) in the box inner top and bottom.
(f) Definition and manufacturing of the electrical circuit to program the textile activation and programming (Fig. 7.58).

An initial phase of activation tests involved the individual analysis of the conductive circuits' properties and the development of an electrical circuit to program and study textile behavior and dynamic light.

With the objective to present and analyze the prototype in a spatial context, a contact was established with Casa da Música (House of Music)—Porto. Casa da Música produces an open house event with an invited country every year, being Great Britain in 2017. A relation between the prototype behavior with the work of a British composer was proposed for the creation of a responsive environment.

Fig. 7.58 Prototype fabrication and testing of the electrical circuit

The connection between sound and visual dimensions were explored through the expressive qualities of the temporal forms.

This section presents the textile behavior and dynamic lighting analysis conducted at House of Music. The variables proposed to explore textile and light expressions focused on color and pattern effects, varying luminosities, scale, direction and rhythms of changes. In addition, parameters of the prototype physical context were also analyzed. Figure 7.59 presents the prototype at a view distance with the overall appearance of the textile light transmittance through the layered morphology and colors along the printed and folded patterns.

In colorized state, the light transmittance pattern is built up by colors composition and folded structure, while during color change activation, the heating lines add a geometric element to the pattern (Fig. 7.60). In the majority of prototype surface, each conductive circuit heats up areas with more than one color, displaying the color ratio from different to similar along the linear forms. Simultaneous color change activation displayed this effect in adjacent or separated areas.

When complete color change was attained, solid color areas were defined with higher luminosities. Figure 7.61 depicts the influence of color lightness in the intensity levels of color change, with similarities in light green and yellow areas (image bottom right) and contrast with darker greens (image top left). The transition to a lighter color has enabled a more obvious distinction between light transmittance in the tabs, which have three pile-up layers and the squares with five. In areas with darker

Fig. 7.59 Dialogues prototype

Fig. 7.60 Prototype section in colorized state (left) and changing color (right)

colors, this effect was barely perceived in particular when observed at a distance, appearing to compose a similar dark pattern (Fig. 7.60 left image).

The linear pattern of color change was also studied according to heating circuits' parameters and the prototype placement. The integration of conductive threads in the weft of the woven substrate was conducted at 1 cm distance each, whereas in

Fig. 7.61 Color change activation with areas with complete decolorization

some areas the connection of two series circuits was intercalated, accomplishing 2 cm distance between each linked thread. Left image of Fig. 7.62 displays a section of the prototype with color change stripes at the two different distances, where a squaretwist area activated with the lower distance lines being in complete colorless. In addition,

Fig. 7.62 Color change linear patterns

the two modules of the prototype presented (M2 and M3) had the same angle rotation of the crease pattern in the weaving plan but the direction of the conductive threads was transformed by the modules vertical and horizontal positioning.

Right image of Fig. 7.62 presents M1 and M2, which crease pattern had different directions and their positioning in two vertical planes also accomplishes different angles, thus affecting the geometry of the color change stripes. As discussed in Sect. 4.2, conductive threads are mostly woven in the weft (Eichhoff et al. 2013; Stoppa and Chiolerio 2014). Textile direction and positioning variables can be explored to design TC textiles with variations in the geometry of the color change pattern.

Shape change of squaretwist morphologies was previously discussed with a light color sample. In this prototype, darker color squaretwists presented dim shades in the square and tabs, which have influenced how movement and shapes were perceived. For example, at a front view, perception of the textile morphology is reduced, namely the folded square depth (Fig. 7.63).

In this prototype, shape change activation also changed the textile color, given all screen printing pastes were elaborated with TC pigments. As a result, heat variation in the Nitinol alloys also introduced a linear color change pattern particularly perceived during fold up at a side view (Fig. 7.64 left image). In fold up and fold down activation cycles, the thickness of color change stripes varied, being able to attain areas with full color change, thus increasing luminosities and the perception of the textile folded morphology (Fig. 7.64 right image).

Textile morphology was influenced by squaretwist direction, particularly in larger shapes, after shape change being activated (Fig. 7.65). In horizontal upside down positioning the squaretwists are small (4 and 6 cm square diagonal) and placement decisions also appeared to have greater relevance after shape change activation.

For the prototype presentation, an interaction diagram was delineated and a framework that relates sound and visual expressions was studied and defined. The interaction diagram was similar to the previous prototype, as presented in Fig. 7.49. In this case, the sensors intended to detect movement by the visitors at a strategic distance

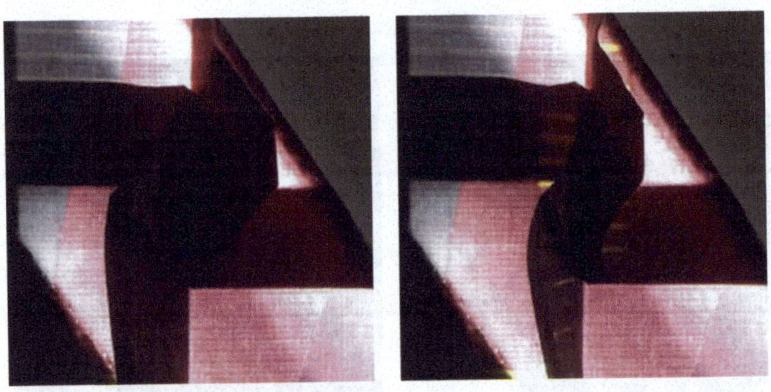

Fig. 7.63 Shape change in a dark color squaretwist

Fig. 7.64 Color change during shape change activation

Fig. 7.65 Shape change

from the prototype, creating possibility to observe the overall dynamic performance from the activation triggering location.

Textile color and shape change activation was inspired by British composer Fredrick Delius' Concerto in C minor. A study was conducted to draw a connection between sound and visual expressions based on intensity levels of change. Excerpts of the musical composition were selected considering three parameters: volume intensity, Low Frequency (LF) and High frequency (HF) response (bass and treble, respectively) and rhythm. These parameters delineated a framework for the textile activation sequence variables.

High intensity sound levels and faster rhythms defined higher number of simultaneous activations of chromic behavior, attaining high luminosity changes. Low volume intensity and slower paces defined more subtle light transmittance variations either with shape change and/or color change in few textile areas.

The relation between HF and LF levels established activation areas. HF levels guided activation definition in the uppermost part of the prototype and LF in the lower part. Level variation defined a direction of the circuits' sequential activation. For example, when the music excerpts had a predominant change from LF to HF, textile activation sequences encompassed a direction of changes from bottom to top areas.

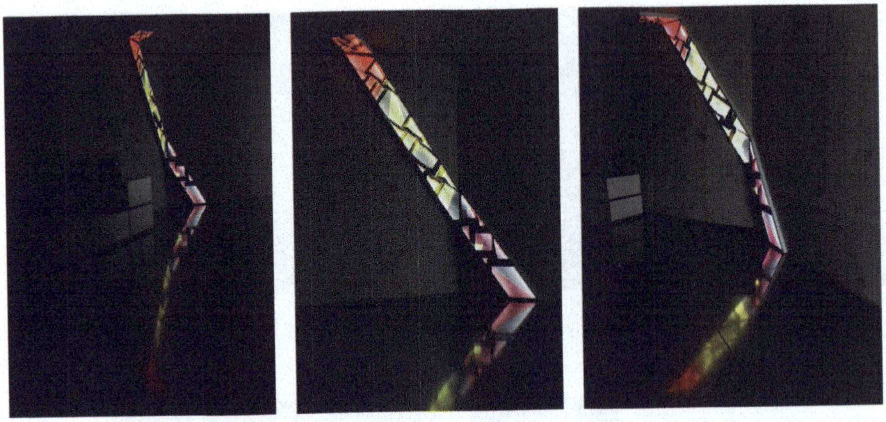

Fig. 7.66 Prototype and light reflection at left, front and right perspectives, respectively

These relations established a conceptual matrix exploring expression of textile behavior and dynamic light. A visual interpretation of the music was not the goal. Furthermore, textile heating and cooling paces were very slow in comparison to the music rhythms.

In the responsive environment created, the beginning of the activation sequences was obviously perceived via sound. Visitors' movement and location in the space triggered the selection of a sequence, which initiated the playing of a music excerpt and the programmed activation. The definition of the interaction parameters has emphasized the link between sound and visual performances, as occurring and stopping at the same time.

Variables of the physical context had played a crucial role on the prototype qualities and how the changes were perceived. The interior space structure has influenced the textile positioning, defining its final configuration in relation to the angle of the vertical planes in M1 and M2, while the ceiling height set the distance of M3 from the floor. In addition, the prototype scale was augmented through the reflex of colored lights on the floor, which geometry depended on the viewpoint. Figure 7.66 displays the left, front and right perspectives, respectively.

This effect has captured attention and movement around the prototype. Depending on the viewpoint and distance, changes could also be perceived through the reflection, more comprehensively with M1 color change, as observed in Fig. 7.67 with the color change pattern in the textile and reflected on the floor.

Exploring color change behavior in association with gradual sound volume variation, differences in the heating circuits connections within consecutive or intercalated conductive threads enabled the creation of diverse expressions. Figure 7.68 depicts a sequence where color change started in a squaretwist area with the heating lines at a lower distance and after approximately 20″ of activation, heating lines in the adjacent areas start to occur simultaneously, with conductive threads at higher distances. The chromic behavior created gradual color and pattern transitions with higher color

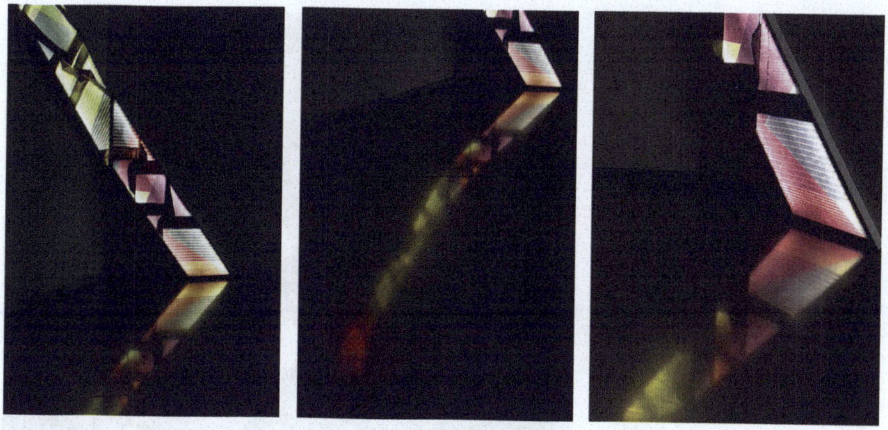

Fig. 7.67 Light reflection of the color change pattern

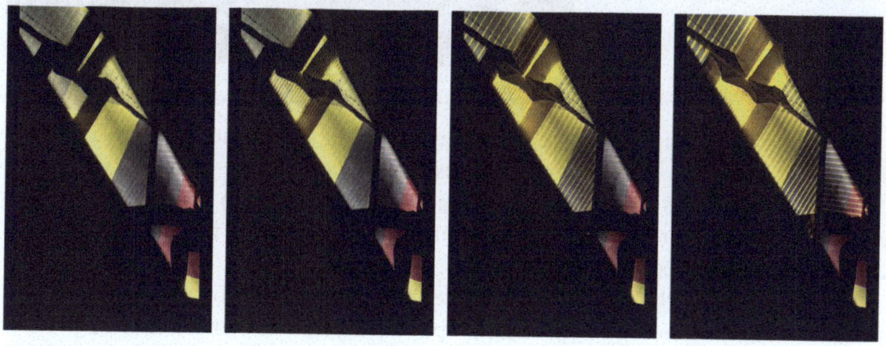

Fig. 7.68 Gradual color and pattern changes

change, attaining more intense luminosities in center area and more disperse at the sides.

Figure 7.69 displays a combination of color change expressions, created with a sequence of color and shape change activations. In the upper area, two squaretwists and tabs changed in a fast pace to full color change, showing obvious brighter areas. In the lower area, color change through heating threads connected with longer distance displayed a striped pattern in the tabs and in areas with one layer, whereas in the squaretwist area, a spotty color change pattern was observed. These expressions transformed light within lower luminosities. In the M1 squaretwist, shape change activation combines movement with the linear chromic pattern. From the photo record viewpoint, both changes were obviously perceived and the change of light transmittance has created a brighter area in the tab backside, whereas the overall light transmittance was low. The expressions analyzed have created a rhythm from higher to lower luminosity changes within a textile direction, occurring from the upper to the bottom areas depicted in the photo.

Fig. 7.69 Dynamic color and pattern expressions

Dynamic possibilities explored with textile color, shape and light have demonstrated the influence of a set of variables in the expressions and intensity levels of change. *Dialogues* prototype established an obvious role in shaping the interior environment and expressively changing ambient light, whereas its dynamic qualities were mostly perceived as subtle and calm.

The geometric light transmittance pattern was build up through different colors and the folded morphology with different pile-up layers. Furthermore, through color change activation, a linear pattern with similar color was also introduced, being influenced by the textile positioning. Through these parameters, intensity levels of change were studied.

The visual complexity of the surface appeared to have turned less obvious in how the chromic changes were perceived. Color and pattern effects were distinguished, although the expressions of change were mostly subtle. The shape change transformation with the squaretwists' tabs rotation about a square diagonal was more clearly identified at a side or perspective view. Observed from the front, light transmittance perception was reduced.

Activation sequences definition was explored through a framework that associated sound and visual expressions. The method used has established a guideline for design decisions on textile behavior activation, namely the definition of simultaneous or sequential activations to explore different intensities of the rhythms and luminosity variations as well as location of changes within the prototype height and changes direction. By focusing on these specific parameters, expression possibilities of textile and light temporal forms were identified and demonstrated.

Physical context parameters were studied through the prototype presentation, namely architectonic structure and scale, as well as surface properties. Expressions of change highlighted have included design possibilities within color change pattern, scale of changes emphasized by light reflection on the floor and differences in how changes are perceived according to the viewpoint.

Designing for subtle or obvious changes involve a complex relation between design variables. The research conducted through the prototypes demonstrated their importance in the design process of textile behavior and dynamic light.

References

Benitez M (2016) Sensorial space: responsive interiors through smart textiles. In: Schneiderman D, Winton A (eds) Textile technology and design: from interior space to outer space. Bloomsbury Academic, New York

Bradley S (2012) 3 types of rhythm you can create visually [Online]. http://vanseodesign.com/web-design/visual-rhythm/. Accessed 2 Oct 2014

Cabral I, Silva C, Worbin L, Souto AP (2017) Exploring dynamic lighting, colour and form with smart textiles. IOP Conf Ser Mater Sci Eng 254:072005. https://doi.org/10.1088/1757-899X/254/7/072005

Clarke S (2011) Textile design. Laurence King, London

Cuttle C (2015) Lighting design: a perception-based approach. Routledge, London and New York

Eichhoff J, Hehl A, Jockenhoevel S, Gries T (2013) Textile fabrication technologies for embedding electronic functions into fibres, yarns and fabrics. In: Kirstein T (ed) Multidisciplinary know-how for smart textiles developers. Woodhead Publishing, Oxford

Hall C, Pederson D, Bryson G (2002) Northern lights: the science, myth, and wonder of aurora borealis. Sasquatch Books, Seatle

Jackson P (2011) Folding techniques for designers: from sheet to form. Laurence King Publishing, London

Jansen B (2013) Composing over time, temporal patterns—in textile design. Doctoral thesis, University of Borås

Le Poidevin R (2015) The experience and perception of time. In: The Standford encyclopedia of philosophy [Online]

Loke L, Reinhardt D, Mcneilly J (2015) Performer-machine scores for choreographing bodies, interaction and kinetic materials. In: Proceedings of the 2nd international workshop on movement and computing 2015 Vancouver. ACM

Mossé A (2016) Self-actuated textiles, interconnectivity, and the design of the home as a more sustainable timescape. In: Schneiderman D, Winton A (eds) Textile technology and design: from interior space to outer space. Bloomsbury Academic, New York

Nefs HT (2008) On the visual appearance of objects. In: Schifferstein H, Hekkert P (eds) Product experience. Elsevier, San Diego, CA

Redström J (2010) On technology as material in design. In: Redström M, Redström J, Mazé R (eds) IT+Textiles. The Interactive Institute and the Swedish School of Textiles, Borås

Rutzky J, Palmer CK (2011) Shadowfolds: surprisingly easy-to-make geometric designs in fabric. Kodansha International, New York

Stoppa M, Chiolerio A (2014) Wearable electronics and smart textiles: a critical review. Sensors 14:11957. https://doi.org/10.3390/s140711957

Vallgårda A (2014) Giving form to computational things: developing a practice of interaction design. Pers Ubiquit Comput 18(3):577–592. https://doi.org/10.1007/s00779-013-0685-8

Weiser M, Brown JS (1995) Designing calm technology [Online]. http://www.ubiq.com/hypertext/weiser/calmtech/calmtech.htm. Accessed 16 Feb 2017

Worbin L (2010) Designing dynamic textile patterns. Doctoral thesis, University of Borås

Yahnin AG, Sergeev VA, Gvozdevsky BB, Vennerstrem S (1997) Magnetospheric source region of discrete auroras inferred from their relationship with isotropy boundaries of energetic particles. Ann Geophys 15:943–958. https://doi.org/10.1007/s00585-997-0943-z

Chapter 8
Conclusions

The design potential of textiles to interact with light through smart materials adaptive and reactive behavior was researched, comprising of two perspectives. A materials research focused on integration processes of the smart and conductive materials in textile substrates and a design research that studied dynamic qualities of textile color and shape change behavior in interaction with light.

Throughout an interdisciplinary approach, the results attained and discussed, built up an understanding on design variables that influence textile and light behavior and their expressions of change and proposed processes to create color and shape thermo-responsive textiles.

Systematic processes of paste recipe elaboration with TC and conventional pigments were developed for the development of screen printed textiles that change color according to a predefined color change ratio, as well as color reproduction of TC textiles.

For textile chromic behavior from different to similar with temperature increase, it was verified that color and concentration of the conventional pigments are fixed parameters and they are variable for TC pigments. For textile that changes from similar to different colors, the process developed applies instrumental methods. The systematic procedures outlined comprises of the development of a database with the colourimetric properties of conventional and TC pigments colors, digital-based formulation of paste recipe considering a pattern color, visual and instrumental assessing of the results and possible adjustments. The results have demonstrated the possibility to create color change effects as predefined, from similar to different with temperature increase.

Conductive threads integration in textiles for electrical activation of color change enables the design of varied chromic effects. Integration pattern of conductive threads in woven structures defines the possible areas of textile heat transfer. The analysis examined that resistive heating is not necessarily homogeneous along the threads length, affecting temperature variation in the textile and respective areas where the TC pigment transits from colorized to decolorized state. Thus, color change patterns also depend on conductive threads' expressive qualities. The outcomes observed

I. D. Cabral et al., *Dynamic Light Filters*, Springer Series in Design and Innovation 3, https://doi.org/10.1007/978-3-030-39529-2_8

were defined as homogeneous or heterogeneous with the latter distinguished between irregular lines and traces.

Moreover, textile heat variation can also change in respect to the substrate thermal conductivity and different contact areas between the conductive threads and the textile substrate, influencing how the textile changes color.

Conductive pigments electrical properties are affected by the printing parameters, whereas resistive heating with the pigment applied demonstrated that the differences were not significant for activation of color change.

In addition, textile folding can vary or damage the materials' electrical conductivity, being an important parameter when designing textiles with color and shape behavior.

The research conducted with shape memory textiles encompassed a novel approach of Nitinol wires being defined accordingly to a textile geometric morphology and integration of Nitinol segments with different shapes in the same woven textiles. The process developed has produced shape memory textiles with the ability to perform two shapes change behavior similar to predefined origami morphologies.

The workflow setup proposed consists of defining the origami morphology and shape change transformation, selecting the Nitinol wire and defining geometries according to their integration in the woven substrate; drawing and production of the dies; optimizing the annealing parameters following the Nitinol heat treatment and finishing processes; and integration of Nitinol wires in the weaving process.

The process outlined the ability to develop textile shape change with one or two different morphologies based on a bias mechanism, considering resistive heating activation method. The relation between shape change transfer of two Nitinol alloys' groups to the host structure at different times suggested that a compromise on the definition of one memorized morphology was required. Designing shape memory textiles involve a set of relationships between Nitinol and textile parameters, namely morphology, weight and positioning.

Research on textile color and shape behavior in interaction with light focused on how a set of design variables can influence the dynamic qualities and expressions of change.

Chromic effect on light transmittance showed ability to change light intensity and tone, in respect to TC pigments color, concentration and application processes in the substrate. With the shape change, textile morphology based on origami tessellation techniques to perform layer number variation created changes in light transmittance intensity and changes of pattern with different luminous on the textile surface.

In working with two thermo-responsive materials in the same textile structure, the designer has to consider thermal variation for both behaviors, according to the materials' activation temperature selected. Possibilities to design simultaneous or independent changes can be explored to extend textile and lighting dynamic qualities.

Temperature as a dynamic design variable encompasses several relationships between the electrical activation definition, physical context and textile characteristics. Studies on activation parameters outlined the importance of electrical current value, constant or variable; activation time and non-activation time. Dynamic behavior of heat sensitive materials encompasses changes in heating and on cooling through gradual transitions. As a consequence, textile changes can also occur after interruption of power.

Room temperature was also found to be a critical aspect, particularly in thermal variation of the textile substrate. In addition to textile material properties and structure, morphology also influences heat variation and with textile origami morphologies, the possibility to design patterns of color change was presented.

Introducing discussion on textile behavior with different intensity levels of change, the research identified and discussed a set of design variables able to influence subtle or evident expressions. Parameters highlighted through the design examples include: color ratio, morphology and transformations type, viewpoint, scale of change, positioning, electrical current values, activation and non-activation time, activation sequences, room temperature, architectonic structure, surface properties. These are tools to explore and design expressions and dynamic qualities of thermo-responsive textile behavior.

Light transmittance variation created through textile color and shape behavior can be defined through qualities such as paces of change, rhythms and expressions of the temporal forms. Activation of diverse sequence configurations enable the creation of dynamic and interactive lighting environments, where dynamic qualities can be perceived as obvious or subtle, according to how textiles and textile behavior are designed.

Bearing in mind a new generation of textile products with new functional and expressive possibilities, Dynamic Light Filters concept can be applied for interior or exterior textile applications that in interaction with light enable to embed dynamic qualities in our environment.

Emergent possibilities of adaptive and reactive behavior of color and shape changing textiles reveal vast potential for the creation of lighting scenarios, encompassing functional and expressive purposes. In interior spaces deprived of natural light, design possibilities within subtle intensity levels of change can establish a sense of rhythm through embedded technology in the textiles. Calm expressions can be explored as an integral and unobtrusive element of the environment, in which dynamic behavior—textiles and light—evolves without necessarily engaging or requiring attention. For building facades, dynamic qualities of smart textiles can be explored within luminous changes, energy issues as well as architectural visual aesthetics.

Appendix A

Results of conductive threads' study presented in Sect. 4.3.1.
Note: In most conductive threads' experiments, three voltage stages were conducted with fifteen seconds each, to attain a maximum temperature below or approximately 45 °C. Conductive threads that present less voltage stages were due: high temperatures detected through the Infrared camera or voltage limit of the power supplied used—30 V.

CT1	V (V)	I (A)	R (Ω)	Tmax.(°C)	IR image	Photo
	1.60	0.16	10.00	34.2		
	1.80	0.18	10.00	35.8		
	2.00	0.19	10.53	38.6		
CT2	**V (V)**	**I (A)**	**R (Ω)**	**Tmax.(°C)**	**IR image**	**Photo**
	0.80	0.28	2.86	36.9		
	1.00	0.31	3.23	40.5		
	1.20	0.38	3.16	41.7		
CT3	**V (V)**	**I (A)**	**R (Ω)**	**Tmax.(°C)**	**IR image**	**Photo**
	10.00	0.01	1000.00	29.7		
	15.00	0.01	1500.00	33.2		
	20.00	0.02	1000.00	42.4		

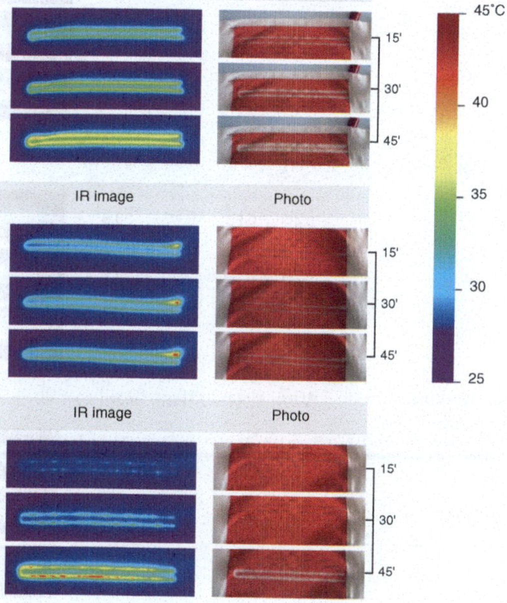

I. D. Cabral et al., *Dynamic Light Filters*, Springer Series in Design and Innovation 3, https://doi.org/10.1007/978-3-030-39529-2

CT4	V (V)	I (A)	R (Ω)	Tmax.(°C)	IR image	Photo
	15.00	0.01	1500.00	30.1		
	20.00	0.01	2000.00	34.4		
	25.00	0.02	1250.00	41.6		

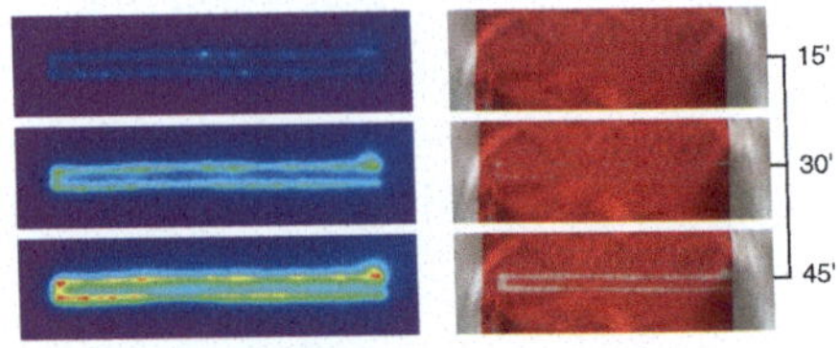

CT5	V (V)	I (A)	R (Ω)	Tmax.(°C)	IR image	Photo
	1.00	0.09	11.11	28.8		
	2.00	0.17	11.76	33.6		
	3.00	0.25	12.00	42.0		

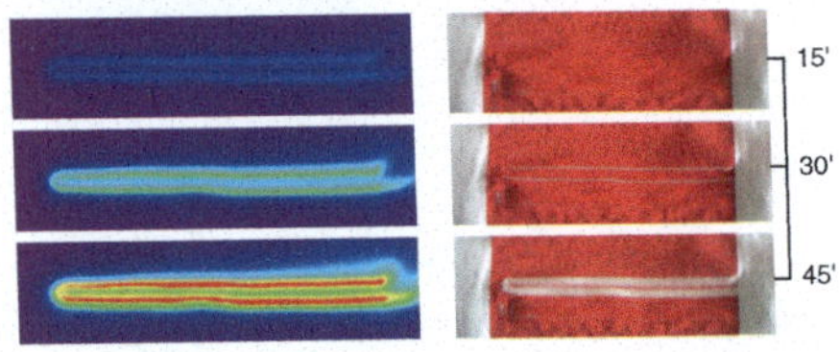

CT6	V (V)	I (A)	R (Ω)	Tmax.(°C)	IR image	Photo
	0.20	2.16	0.09	33.2		

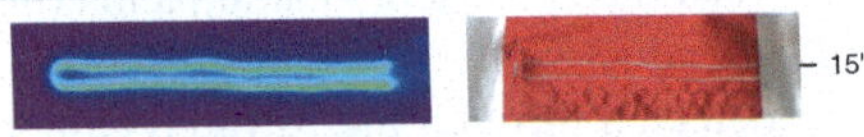

CT7	V (V)	I (A)	R (Ω)	Tmax.(°C)	IR image	Photo
	0.30	0.56	0.54	32.2		
	0.40	0.73	0.55	35.1		
	0.50	0.87	0.57	40.1		

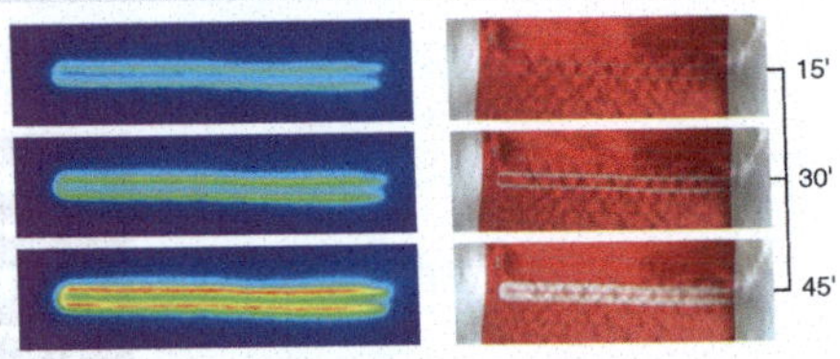

CT8	V (V)	I (A)	R (Ω)	Tmax.(°C)	IR image	Photo
	0.40	0.60	0.67	33.8		
	0.50	0.70	0.70	37.0		
	0.60	0.78	0.70	40.3		

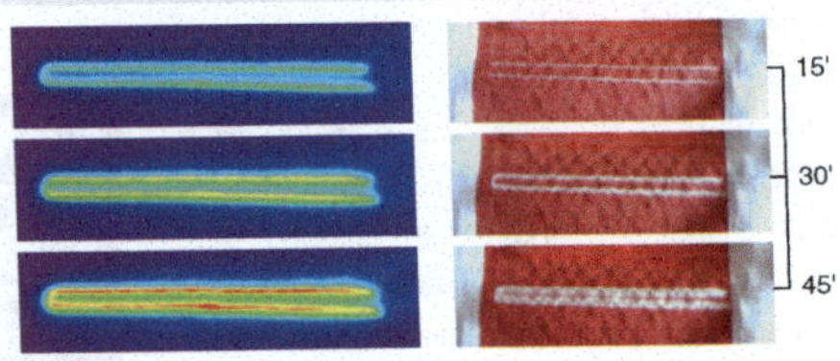

CT9	V (V)	I (A)	R (Ω)	Tmax.(°C)	IR image	Photo
	2.00	0.11	18.18	32.7		
	2.50	0.13	19.23	35.6		
	3.00	0.16	18.75	39.5		

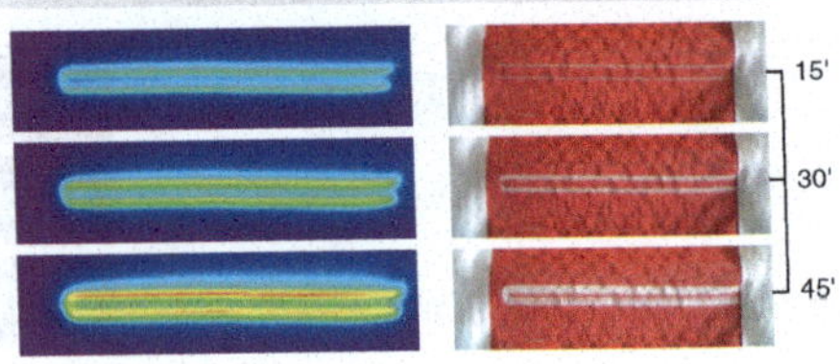

CT10	V (V)	I (A)	R (Ω)	Tmax.(˚C)	IR image	Photo
	1.40	0.13	10.77	35.0		
	1.60	0.16	10.00	37.0		
	1.80	0.18	10.00	40.4		

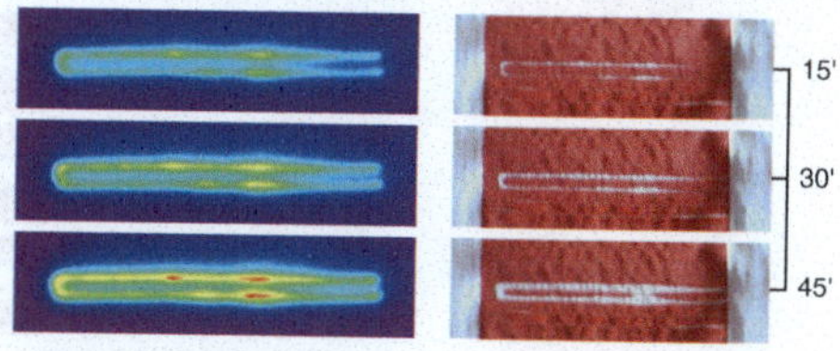

CT11	V (V)	I (A)	R (Ω)	Tmax.(˚C)	IR image	Photo
	1.00	0.63	1.59	38.5		

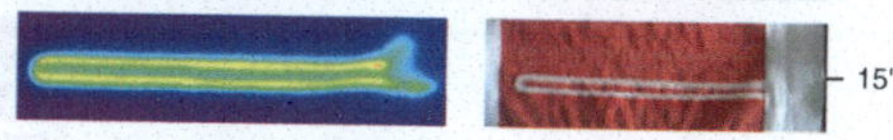

CT12	V (V)	I (A)	R (Ω)	Tmax.(˚C)	IR image	Photo
	1.00	0.17	5.88	32.4		
	1.50	0.25	6.00	38.6		
	2.00	0.33	6.06	48.1		

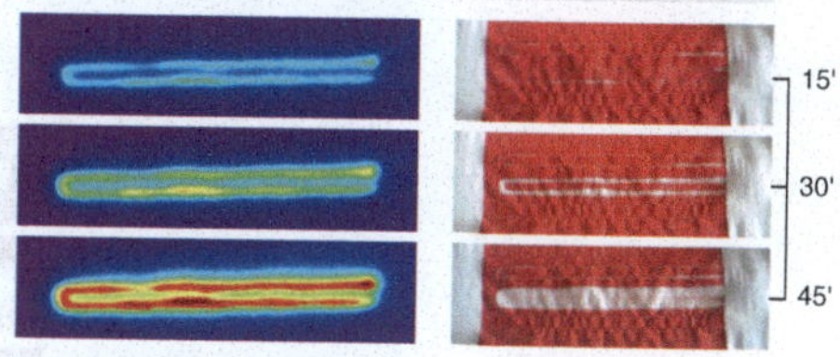

CT13	V (V)	I (A)	R (Ω)	Tmax.(˚C)	IR image	Photo
	1.50	0.06	25.00	32.4		
	2.00	0.08	25.00	38.0		
	2.50	0.11	22.73	43.3		

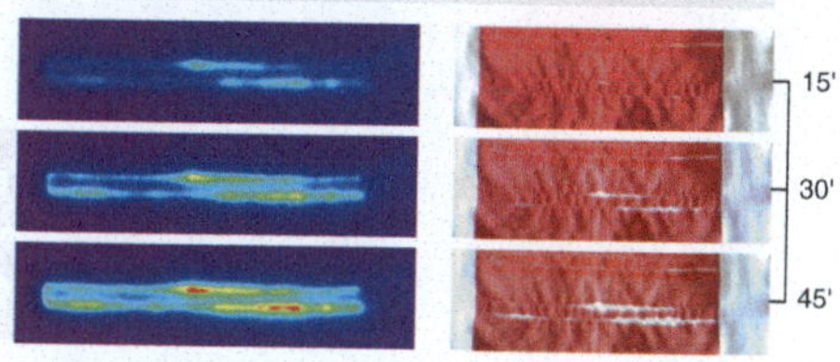

CT14	V (V)	I (A)	R (Ω)	Tmax.(˚C)	IR image	Photo
	4.00	0.08	50.00	38.9		
	5.00	0.11	45.45	46.2		

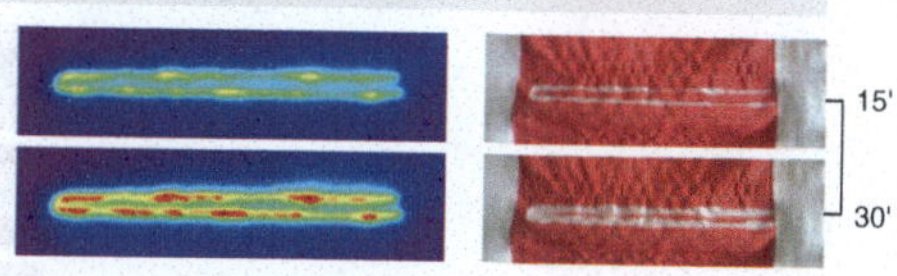

CT15	V (V)	I (A)	R (Ω)	Tmax.(˚C)	IR image	Photo
	4.00	0.02	200.00	37.4		
	5.00	0.03	166.67	42.0		
	6.00	0.04	150.00	49.2		

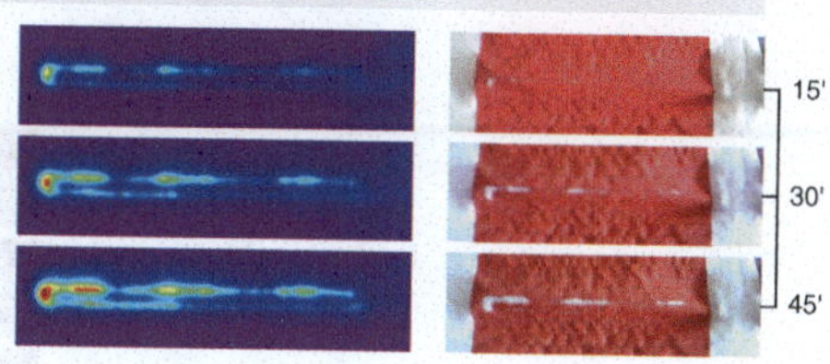

CT16	V (V)	I (A)	R (Ω)	Tmax.(˚C)	IR image	Photo
	8.00	0.03	266.67	34.2		
	9.00	0.03	300.00	36.3		
	10.00	0.04	250.00	38.9		

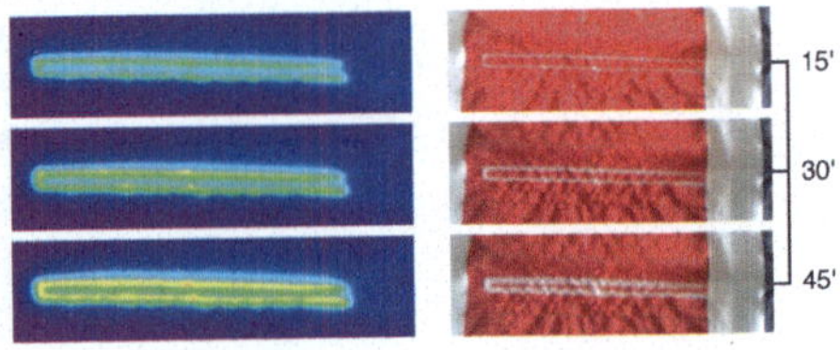

CT17	V (V)	I (A)	R (Ω)	Tmax.(˚C)	IR image	Photo
	4.00	0.08	50.00	35.3		
	5.00	0.10	50.00	40.4		

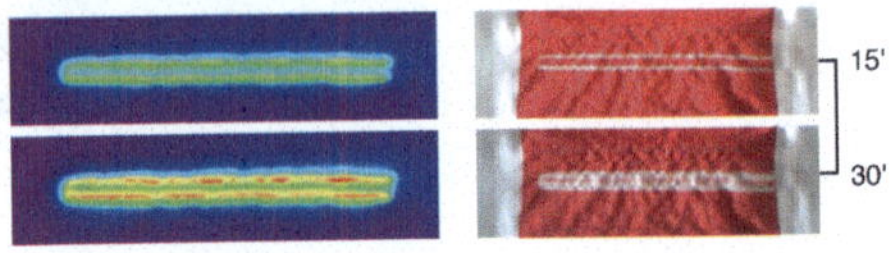

CT18	V (V)	I (A)	R (Ω)	Tmax.(˚C)	IR image	Photo
	20.00	0.01	2000.00	37.1		
	25.00	0.01	2500.00	41.4		
	30.00	0.01	3000.00	45.7		

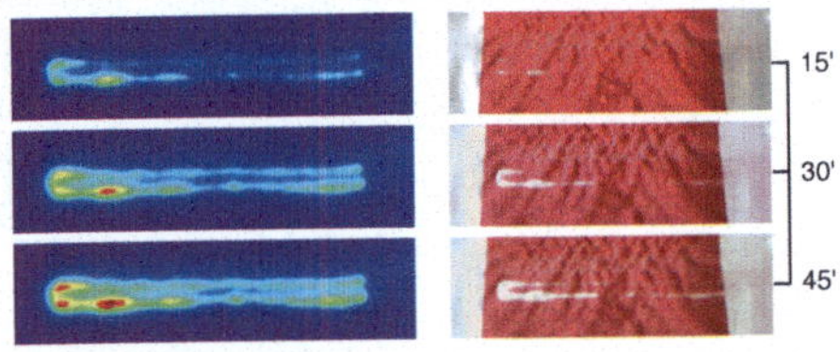

CT19	V (V)	I (A)	R (Ω)	Tmax.(˚C)	IR image	Photo
	3.00	0.11	27.27	37.7		
	4.00	0.15	26.67	47.3		

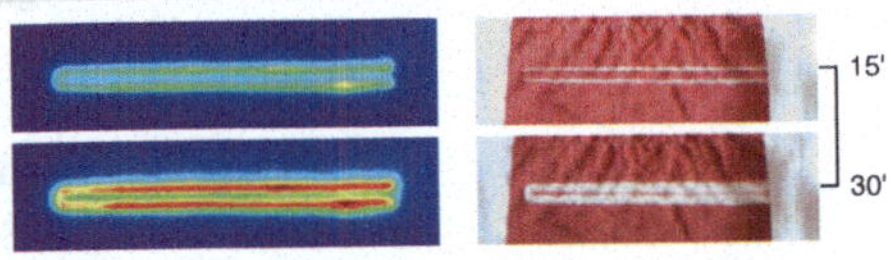

CT20	V (V)	I (A)	R (Ω)	Tmax.(˚C)	IR image	Photo
	6.00	0.04	150.00	33.5		
	7.00	0.05	140.00	36.1		
	8.00	0.06	133.33	39.5		

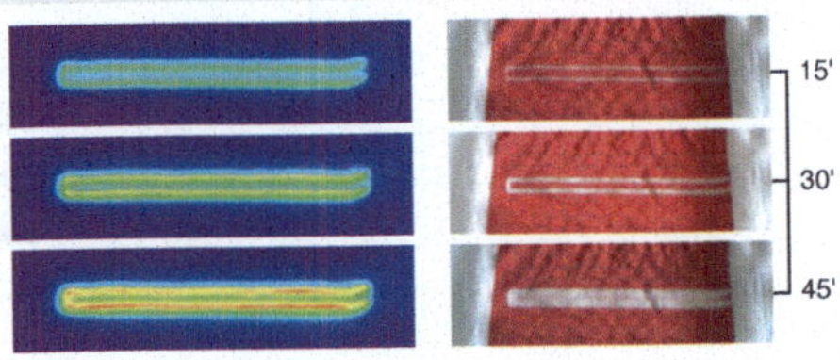

CT21	V (V)	I (A)	R (Ω)	Tmax.(˚C)	IR image	Photo
	4.00	0.04	100.00	34.1		
	5.00	0.04	125.00	37.7		
	6.00	0.05	120.00	43.4		

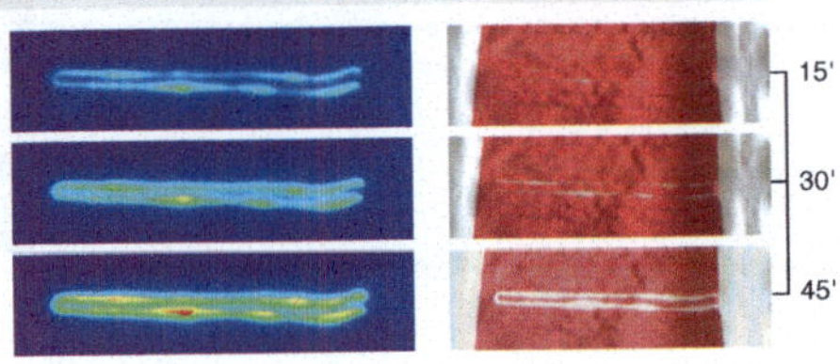

CT22	V (V)	I (A)	R (Ω)	Tmax.(°C)
	2.00	0.04	50.00	34.6
	3.00	0.07	42.86	44.9

IR image / Photo

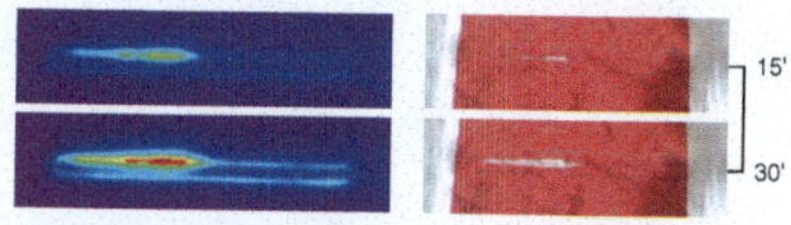

CT23	V (V)	I (A)	R (Ω)	Tmax.(°C)
	0.50	0.60	0.83	34.5
	0.70	0.78	0.90	42.2

IR image / Photo

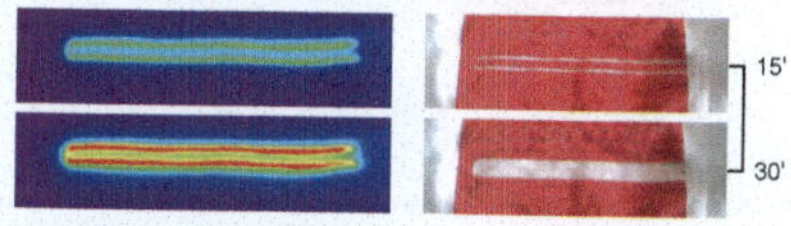

CT24	V (V)	I (A)	R (Ω)	Tmax.(°C)
	0.30	1.03	0.29	34.6
	0.40	1.43	0.28	42.9

IR image / Photo

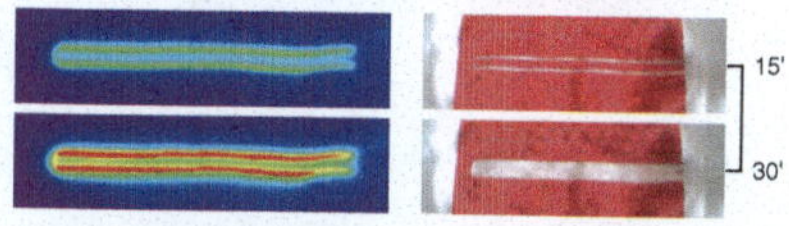

CT25	V (V)	I (A)	R (Ω)	Tmax.(°C)
	8.00	0.03	266.67	33.5
	10.00	0.04	250.00	38.0
	12.00	0.04	300.00	43.0

IR image / Photo

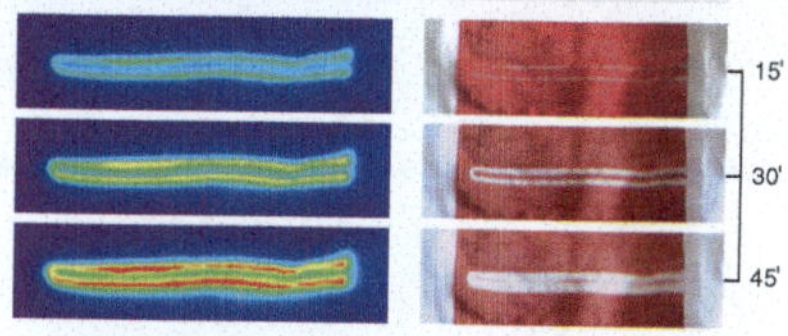

CT26	V (V)	I (A)	R (Ω)	Tmax.(°C)
	1.00	0.21	4.76	32.7
	1.20	0.24	5.00	35.7
	1.50	0.30	5.00	41.0

IR image / Photo

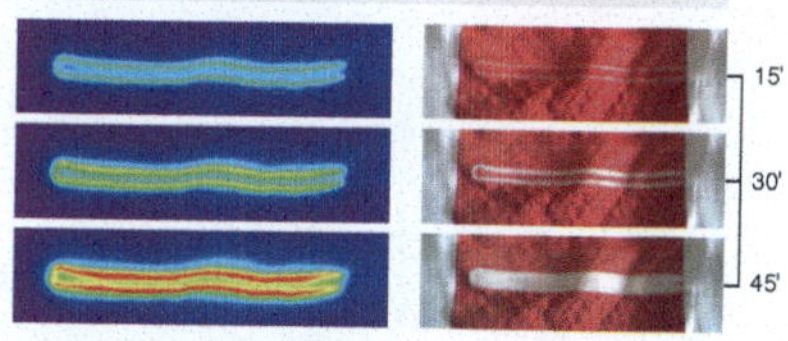

CT27	V (V)	I (A)	R (Ω)	Tmax.(°C)
	7.00	0.04	175.00	34.2
	8.00	0.05	160.00	39.3
	9.00	0.06	150.00	44.2

IR image / Photo

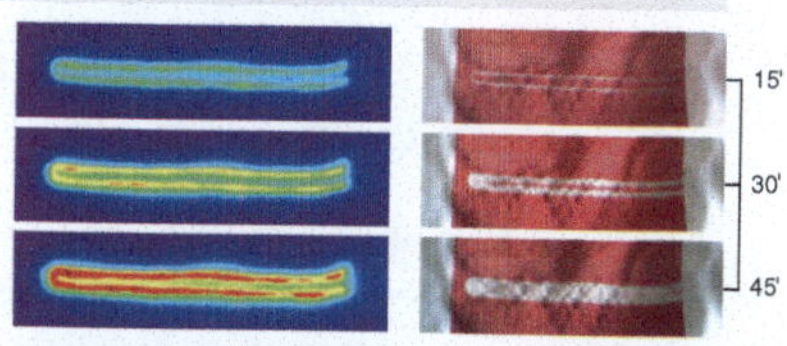

CT28	V (V)	I (A)	R (Ω)	Tmax.(°C)
	5.00	0.05	100.00	33.6
	6.00	0.06	100.00	37.2
	7.00	0.07	100.00	41.3

IR image / Photo

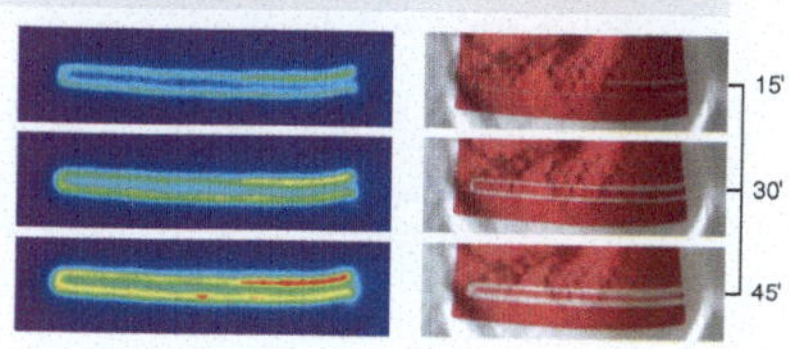

Appendix B

Description of samples X and Y used in experimental work presented in Sect. 7.1.1. Sample X was used in the color change tests A to D and sample Y in test E. They were woven in the loom Jacquard Vamatex with 41.2 tex CO warp and 14.7 tex PES weft, integrating the conductive thread Karl Grimm High Flex 3981 in the weft of a plain weave structure at 1 cm distance each (13 picks PES and 1 pick conductive thread per cm). Conductive threads insertion in the weaving was programmed through the loom software.

With objective to connect the conductive threads into a series circuit, a complementary sample was developed for analysis of three connection alternatives: silver solder, crimp bead and double knot. The soldering process appeared to have weakened the conductive threads and during the textile handling and power supply, some connections broken near the solder point. Both crimp bead and double knot appeared stable, they did not overheat during power supply and electrical resistance of the threads connected by each process were similar. With crimp bead connection, a slightly flatter shape was attained and it was relatively faster to do, being the connection alternative selected.

Previously to the conductive threads connection, samples X and Y were screen-printed with 10% TC red. Additionally, sample Y was folded with a squaretwist pattern.

I. D. Cabral et al., *Dynamic Light Filters*, Springer Series in Design and Innovation 3, https://doi.org/10.1007/978-3-030-39529-2

The manufacturer's authorised representative in the EU is Springer Nature Customer Service Centre GmbH, Europaplatz 3, 69115 Heidelberg, Germany. If you have any concerns regarding our products, please contact ProductSafety@springernature.com

Printed and bound by CPI Group (UK) Ltd, Croydon, CR0 4YY
22/09/2026
02221060-0005